TOURISM AND SUSTAINABLE ECONOMIC DEVELOPMENT

TOURISM AND SUSTAINABLE ECONOMIC DEVELOPMENT

edited by

Amedeo Fossati
Department of Economics and Fiscal Studies
University of Genova, Italy

and

Giorgio Panella
Department of Public Economics
University of Pavia, Italy

Kluwer Academic Publishers
Boston / Dordrecht / London

Distributors for North, Central and South America:
Kluwer Academic Publishers
101 Philip Drive
Assinippi Park
Norwell, Massachusetts 02061 USA
Telephone (781) 871-6600
Fax (781) 871-6528
E-Mail < kluwer@wkap.com >

Distributors for all other countries:
Kluwer Academic Publishers Group
Distribution Centre
Post Office Box 322
3300 AH Dordrecht, THE NETHERLANDS
Telephone 31 78 6392 392
Fax 31 78 6546 474
E-Mail < orderdept@wkap.nl >

 Electronic Services < http://www.wkap.nl >

Library of Congress Cataloging-in-Publication Data
Tourism and sustainable economic development / edited by Amedeo Fossati and Giorgio
Panella.   p.cm.
   "Papers presented at a conference on Tourism and sustainable economic development
held in Imperia, Italy, on 29th May 1998" -- Pref.
 Includes bibliographical references.
 ISBN 0-7923-7908-X (alk.paper)
   1. Tourism--Economic aspects--Congresses. 2. Sustainable development-Congresses.
I. Fossati, Amedeo, 1937-II. Panella, Giorgio.

G155.A1 T58938 2000
338.4'791--dc21

00-055992

CONTENTS

LIST OF CONTRIBUTORS

Magda ANTONIOLI CORIGLIANO
Bocconi University, Italy
e-mail: magda.antonioli@uni-bocconi.it

François BONNIEUX
INRA, Rennes, France
e-mail: bonnieux@roazhon.inra.fr

Amedeo FOSSATI
University of Genova, Italy
e-mail: fossati@economia.unige.it

Alessandro LANZA
International Energy Agency, Paris, France, and University College, London
e-mail: lanza@iea.org

Franco LOSURDO
University of Bari, Italy
e-mail: f.losurdo@lex.uniba.it

David MADDISON
Centre for Social and Economic Research on the Global Environment
University College, London
e-mail: d.maddison@ucl.ac.uk

Anil MARKANDYA
University of Bath, United Kingdom
e-mail: hssam@bath.ac.uk

Gianfranco MOSSETTO
University Ca' Foscari, Venezia, Italy
e-mail: nexa@doge.it

Peter NIJKAMP
Free University, Amsterdam, The Netherlands
e-mail: pnijkamp@econ.vu.nl

Giorgio PANELLA
University of Pavia, Italy
e-mail: panella@unipv.it

Francesco PIGLIARU
University of Cagliari, Italy and University of London, U.K.
e-mail: pigliaru@unica.it; fp@soas.ac.uk

Pierre RAINELLI
INRA Unité d'économie et sociologie rurales, Rennes, France
e-mail: rainelli@roazhon.inra.fr

Vincent RENARD
Ecole Polytechnique, Centre Nationale de Recherche Scientifique,
Paris, France
e-mail: renard@poly.polytechnique.fr

Antonio Paolo RUSSO
Erasmus University, Rotterdam, The Netherlands
e-mail: russo@few.eur.nl

Jan Van Der BORG
Ca' Foscari University,Venezia, Italy
e-mail: vdborg@helios.unive.it

INTRODUCTION

For many countries tourism is an industry of great economic significance; tourism is seen as a main instrument for regional development, as it stimulates new economic activities. Tourism may have a positive economic impact on the balance of payments, on employment, on gross income and production, but it may also have negative effects, particularly on the environment. Unplanned and uncontrolled tourism growth can result in such a deterioration of the environment that tourist growth can be compromised: we are left with the phenomenon known as "tourism destroys tourism". The environment, being the major source of tourist products, should therefore be protected in order to have further growth of tourism and economic development in the future: sustainable tourism and sustainable economic development should be reflected upon.

Questions arise as to whether it is possible to keep on developing tourism in a certain area without negative or irreversible influences on the environment. Should we promote tourism and in so doing accept a certain degree of environmental deterioration, or should we give priority to environmental protection and accept less revenue from tourism? The answer hinges on the links between tourism and the environment and vice versa. A good deal of tourism relies upon resources or assets that either cannot be reproduced or cannot easily be reproduced. This is especially true with regard to tourism based on the natural environment as well as on historical-cultural heritage.

This book provides a theoretical framework for these problems, as well as practical illustrations on the following topics:

- the conditions under which specialization in tourism is not harmful for economic growth;
- the trade-offs, if any, between tourism development and economic growth;
- the need for government intervention and the various policy options and instruments available to policy makers.

The book comprises two parts. The first part presents general views on tourism and sustainable economic development, and some opinions on the relationship between tourism and the environment. Some of the basic concepts implicit in sustainability are examined in relation to regional development, urban tourism, art cities and rural tourism. The second part concentrates on strategies and policy instruments.

In the first chapter, Amedeo Fossati and Giorgio Panella describe the positive and adverse effects of tourism activities and the conditions under which sustainable development can be achieved. In defining these conditions economists have found it useful to differentiate between strong and weak sustainability. Strong sustainability tends to stress the limits to substitutability because of the importance of irreversibility, particularly with regard to certain critical aspects of natural capital. Weak sustainability, on the other hand, requires only that the total capital stock be maintained, allowing substitution between man-made and natural components. This distinction is important in order to define the strategies to apply in relation to tourism and economic development. The transmission to future generations of a quantity of natural capital that provides a sustained yield of economic and environmental services, including amenities, is relevant. Translating sustainability goals into action necessitates changing economic policies to maintain natural capital.

The subject has been further developed by Peter Nijkamp. He argues that a balance has to be found between economic growth and ecological preservation of tourist areas. The level of tourist expenditures is to be evaluated against the social costs of the sector concerned. He offers some numerical results on a meta-analytical experiment on tourist income multipliers in various tourist areas.

The role of natural resources as a main factor of economic development has been further analysed by Alessandro Lanza and Francesco Pigliaru. Cross-country data for 1985-95 on tourism specialisation and economic growth reveal the following data: i) many tourism countries have grown faster compared to other countries; and ii) they are small. A two-sector endogenous growth model to obtain explanatory hypotheses about these findings has been used. In particular, the conditions required for small countries to specialise in tourism and enter a faster growth path have been analysed. The suggestion is that what matters is a country's relative endowment of a natural resource, rather than its absolute size.

This topic has also been analysed in relation to cultural tourism and art cities. The contribution by Paolo Russo and Jan van den Borg presents some guidelines for the management of tourism development in heritage cities in the context of sustainable tourism. It is argued that the organisation of the cultural sector and the quality of intersectorial links represent key organisational issues for these cities to find a place on the map of the new economic geography of Europe. At the same time, they provide a fascinating way to escape the "vicious circle" of tourism specialisation. This policy change requires an integral approach to cultural planning, with several actors being involved in its management. Examples of such policies are given for the cities of Antwerp and Venice. The Belgian city has proved successful in combining a strategy of industrial recovery with a new impulse to the tourism sector, while in the Italian town the re-organisation of the cultural system faced with the massive pressure of tourist flows has just begun to gain consensus and strength.

Gianfranco Mossetto also turns his attention to the management of art cities. The cities of art may be identified according to the specific link between their culture (and art) and their economy. One can thus consider "economy-dependent" and "culture-dependent" models, and among these, expansion and decadent (or implosive) models. The management of these cities, and therefore, their development and decongestion policies have different consequences in each situation. They may even have accelerated decadence in a context of harmful dependence of the city economy on its past culture. The usual subsidisation, as practised in Italy, proves to be a very limited and indeterminate tool.

The need for environmental goods is examined in relation to rural tourism by François Bonnieux and Pierre Rainelli. The chapter, which is the last in the first part of the book, draws attention to the relationship between rural tourism and sustainable agriculture since the major part of rural amenities are provided by agriculture. The chapter provides a general framework to assess rural amenities according to their nature and how they can be captured by local communities. An illustrative example deals with sport-fishing in Lower Normandy.

The second part of the book concentrates on strategies and policy instruments. The purpose is to concisely define and bring together some policies which appear to be necessary, and whose implementation is required if we are to reconcile tourism development with the protection and conservation of the environment. Some analytical tools for policy making with regard to tourism and the environment are developed. As will become clear, there are many gaps in our knowledge that need to be filled if we are to be successful in controlling tourism in a way that puts this important industry onto a sustainable development path.

The chapter by Anil Markandya is an attempt to better understand the kind of instruments to be used, how market structure is a relevant factor in determining the levels at which controls are set, and the key parameters that determine the levels of the market based instruments. He discusses the main sources of externality in tourism and the lack of empirical estimates of parameters that measure such external effects. He also outlines some of the important stylised facts about the tourism industry, and provides a discussion on the private market equilibria for such an industry including the relation between such equilibria and the social optimum. Then, some simulations for an industry that supplies tourism services under monopolistic and monopolistically competitive conditions are reported. Finally, some further developments that need to be made in order to analyse more complex and more interesting conditions are discussed.

A tool often suggested nowadays to internalise externalities efficiently is the definition of property rights. A right is a permit from a government or public authority to perform actions. This instrument has strong proponents and opponents; it is important to carry out a careful analysis of the achievements of tradable rights in their specific objectives and social context. In chapter 8, Vincent Renard reviews the property rights approach to land management, and examines how this approach has been used in different countries.

Further instruments and policy options have been examined by Magda Antonioli Corigliano. She states that, apart from providing environmentally-oriented suggestions for territorial planning and taking into consideration the interactions with different types of tourism, it is important to consider the concept of area quality (or product area) referred to a specific tourist destination. Since a tourist product is the result of the co-production of different actors, it mainly consists of immaterial components. The tourist product is often determined by trust and experience and it is perceived and assessed by the single tourist comprehensively. Thus, the tourist destination is the ideal level to define and control a quality process as it is closer to the tourist's experience. The quality strategy of the "tourist district" is achieved by adopting and implementing a Quality Plan. It is paramount that the different suppliers adopt joint and co-ordinated actions, as operators do not necessarily achieve homogenous results.

In chapter 10, Franco Losurdo analyses the problem of coordinating the different subjects and the projects defined at local level. He starts from the concept that sustainable tourism development is an objective that should be territorialised in order to be pursued in practice. As carrying capacity varies considerably at local level, a model of sustainable development should be defined taking into consideration local conditions. Territorial pacts can be used for this purpose. Important are the conditions under which a territorial pact may be a useful instrument for a sustainable tourism development policy.

Finally, in the last chapter David Maddison investigates the impact of climate change on the chosen destinations of British tourists. Destinations are characterised in terms of "attractors" including climate variables, travel and accommodation costs. These and other variables are used to explain the current observed pattern of overseas travel in terms of a model based upon the idea of utility maximisation. The approach permits the trade-offs between climate and holiday expenditure to be analysed, and effectively identifies the "optimal" climate for generating tourism. The findings are used to predict the impact of various change scenarios on popular tourist destinations.

PREFACE AND ACKNOWLEDEGMENTS

This book publishes a selection of the papers and communications presented at the conference on Tourism and Sustainable Economic Development held in Imperia, Italy, on 29th May 1998, organised by the Institute of Public Finance of the University of Genova and the Polo Universitario Imperiese.

The conference was funded by the Fondazione Cassa di Risparmio di Genova ed Imperia, the Amministrazione Provinciale di Imperia and the Casinò of Sanremo. The editors are particularly grateful to these institutions. They would also like to express their appreciation to Bernadette Burke for revising the English language version.

PART I

TOURISM AND SUSTAINABLE DEVELOPMENT: THEORY AND PLANNING

1. TOURISM AND SUSTAINABLE DEVELOPMENT: A THEORETICAL FRAMEWORK

AMEDEO FOSSATI AND GIORGIO PANELLA

1. QUALITATIVE AND QUANTITATIVE ASPECTS

Everyone knows what tourism is and who tourists are: after all, if once we were "todos caballeros", today we can say we are "todos turistas". Nevertheless, scientific analysis allows us to better understand this phenomenon and its implications; thus this report seeks to stimulate reflection on how to better approach the tourist phenomenon and to discuss its interrelations with public intervention.

At the cost of partially distancing ourselves from reality, the scientific approach is based on suitable stylized definitions, so that the reasonings that lead us toward statements can be rigorously or scientifically characterized. In this way we can produce reliable statements, even under the assumptions resulting from the stylized approach. Thus for tourism as well different definitions are adopted according to the aims of the analysis. However, there are three elements that are found in almost all the definitions, and which appear to be sufficient here to provide an analytical framework for our purposes. These elements are:

1. the idea of movement (trip, change of one's normal residence, or one's normal habitat);
2. the idea of a person's behaviour as a tourist (the economic-social role temporarily assumed);
3. the environment, or tourist space (the territory's tourist vocation).

As far as the travel aspect is concerned, it does not seem appropriate to focus here on the fact that travel takes place (motivational point of view: recreation, culture, work, etc.), nor on the economic-operative aspects of travel. With regard to public interventions, it is enough to view the physical movement (trip) as a *de facto* assumption of tourism.

As concerns behaviour, we will have the occasion to return to the social impact of tourism; but for now we will only point out that from the economic point of view we often speak of the *market*, and the *tourist industry or sector*. Beyond the possible schematizations based on demand and supply, we must remember that tourism is quite complex: properly speaking, it is neither a market nor an industry, since tourists are economic agents that operate in many markets in the country they travel to (in the extreme case in all markets). As a result all the industries in the country of destination can produce, or contribute to producing, the goods and services demanded by tourists, while it is difficult to distinguish the part of demand caused by the flow of tourists from that owed to the native population. In this way we often speak of "tourist system"[1]; in turn, the "tourist product" is generally a mere synthetic abstraction: more precisely, a composite product, at most a specific vector of final goods that varies over time and space. It is nevertheless true that there are industries or sectors which are directly and/or intensely concerned with tourism, while others are only indirectly and/or marginally so. In particular the sectors of transport, lodging and catering, and those of tour operators and travel agencies are directly involved in the supply of tourist services, in a prominent way.

This naturally implies that the services supplied to tourists in the host country increase national income, which makes tourism of particular interest for the national objective of economic growth. Moreover, since tourist services are generally *labour intensive*, tourist development tends to increase employment more than proportionately with respect to national income; this is especially important as far as the country's employment objective is concerned. Lastly, we must point out that countries with a vocation for tourism obtain balance of payments benefits, given the inflow of currencies connected with international tourism.

Finally, as regards the territory as a tourist area, the climatic and landscape features (sun, sea, mountains, panoramas) are prominent in a quantitative sense, in that the myth of heat and sun is still important today with regard to a country's tourist vocation. Nevertheless, there is no single "tourism" but many *tourisms* in accordance with the principal motivation: recreation, vacation, health, culture, adventure, congresses, religion, etc.; as a result the features of a territory that can attract tourists are extremely varied. Even an area which is by nature inhospitable (desert, arctic) can attract extreme or adventure tourism. Purely geographic features are important only in a specific socio-cultural context; we can state that the tourist vocation of a territory represents solely an anthropological product: at most any territory can become *touristic* if an activity or a manufactured good is introduced which arouses the interest of potential tourists. Thus the role which public measures can play in this regard is clearly evident, in strengthening and "inventing" the territory's tourist vocation, above all with regard to measures at the local level.

In short, from the qualitative point of view we can conclude that tourist activity is essential for its direct effects on the social, cultural, educational and economic sectors of individual countries, and on international relations throughout the world[2], and as such it is of primary interest to the collectivity and justifies an important public intervention. Before going further, it is appropriate to briefly point out the quantitative dimensions of tourism in order to provide a framework for the possible implications of public intervention.

Tourism is usually said to be the most important industry at world level; it is estimated to represent between 6% and 7% of GDP. For Italy, recent estimates put the value added of the tourist sector at 5.9% of GDP, while tourist expenditures account for 10% of final consumption. The percentage of GDP accounted for by tourism rises to 12.4% if we consider the enlarged sector that contains all the active agents (tourist, firms, public authorities), and to 20% if we also include infrastructure investments[3]. It is true that these estimates should be cautiously considered given the elusiveness of the tourist sector, which was mentioned above. But it is also true that at the world level tourism is in full development, so much so that the World Tourist Organization estimates an average annual increase of more than 3.5% in the international flow of tourists over the next twenty years.

Certain tourist indicators are conventionally used to measure tourist flows[4]; the main problem here is the lack of perfect international homogeneity. In all ways mobility is captured by *arrivals* and *presences,* which in turn are divided into national and foreign. The arrivals are the number of visitors who reach a certain destination; presences are the number of nights spent by the tourists at that destination. Also of interest is the tourist balance (currency) and various indicators of the capacity to receive tourists and the degree of utilization.

From the point of view of tourism, Europe is the most frequented area at the world level, with 59.22% of international tourist arrivals in 1996, followed by the Americas (19.32%) and East Asia and the Pacific rim

(14.65%). Table 1.1 provides data on world tourist flows, which are not only substantial but strongly on the rise, and in any case to an extent greater than national income.

The corresponding currency earnings for 1996 break down as follows: Europe 51%, the Americas 25.1%, East Asia and the Pacific Rim 19.1%, Africa and the Middle East 1.9%, and South Asia 0.9%.

Within Europe the breakdown for foreign tourist arrivals by area and the corresponding currency inflows is presented in Table 1.2, where we see among other things that in East Europe the cost for tourists is much lower than in the rest of Europe.

Table 1.1. International tourist arrivals by geographic area (in millions)

	1992	1993	1994	1995	1996
Africa	17.6	18.0	18.0	19.2	20.6
The Americas	103.4	103.7	106.3	110.4	114.7
East Asia and Pacific Rim	62.7	69.6	76.9	79.7	87.6
South Asia	3.6	3.6	3.9	4.3	4.5
Middle East	8.6	9.0	9.9	13.7	15.3
Europe	307.2	313.7	329.8	336.4	351.6
Total	503.1	517.6	544.8	563.7	593.7

Source: WTO

Moreover, 72% of international tourist flows in 1996 were directed toward only 20 countries, while 71.91% came from only 20 countries. Table 1.3 shows the top six countries in the world in this regard.

Table 1.2. International tourists in Europe in 1996: arrivals and foreign currency inflows (Percentage)

	Arrivals	Foreign currency inflows
Central-Eastern	23.74	11.12
North	11.18	15.81
South	28.02	31.42
Western	33.63	36.71
Eastern Mediterranean	3.43	4.94
Total for Europe	100.00	100.00

Source: WTO

In Italy in 1996, 32.853 million foreigners arrived, while for 1997 and 1998 the estimates are 32.9 and 34 million, respectively. Italy is thus fourth in the world as a tourist country, both from the point of view of arrivals and that of foreign currency earnings (that is, foreign tourist spending in Italy).

Table 1.3. Classification of the most important countries for tourist exports and imports
1996 – Percentage with respect to world flow

Arrivals		Foreign currency earnings		Expenditures on foreign trip	
France	0.10%	U.S.	0.15%	Germany	0.14%
U.S.	0.08%	Spain	0.07%	U.S.	0.13%
Spain	0.07%	France	0.07%	Japan	0.11%
Italy	0.06%	Italy	0.06%	Great Britain	0.07%
Great Britain	0.04%	Great Britain	0.05%	France	0.05%
China	0.04%	Germany	0.04%	Italy	0.04%

Source: WTO; *Note:* Expenditures are for 1995

But it is also sixth in the world in terms of the starting off point for
tourists. Germany, on the other hand, is first in terms of generating tourist
flows, and sixth in the world as a producer of tourism, at least from the
foreign currency point of view.

Tourism is also an internal phenomenon, in addition to being an
international one; thus in Italy the internal flow is greater than that of
foreigners. In fact, in 1996 the total number of arrivals in accommodation
facilities was 69.44 million, 58% of which owes to Italy. Among the most
active tourist regions in Italy are, in order of importance, Veneto (13.9% of
arrivals), Tuscany (12.2%), Lazio (10.7%), Lombardy (11.1%), Emilia-
Romagna (9.5%), Trentino Alto Adige (8.8%), Campania (5.5%), Liguria
(4.8%), and so on for the other regions. Nevertheless, from the point of view
of arrivals with respect to residents, Valle d'Aosta is in first place, followed
by Trentino Alto Adige, Tuscany, Veneto, Umbria, Liguria, Emilia-
Romagna, and Lazio. Finally, from the point of view of arrivals per km²,
Liguria is first, followed by Veneto, Trentino Alto Adige, Lazio, Tuscany,
and Lombardy.

Table 1.4. Overall arrivals for accommodation facilities according to type of tourist
locality; 1995 – Percentages

	Italians	Foreigners	Total
Art Cities	22.1	38.9	29.0
Mountains	10.1	5.4	8.2
Lakes	3.5	8.4	5.5
Sea-costs	27.7	19.8	24.5
Spas	3.5	3.5	3.5
Hillsides	3.3	2.6	3.0
Chief Provincial towns	8.9	8.3	8.7
Others	20.9	13.1	17.7
Total	100.0	100.0	100.0

Source: ISTAT

2. EXTERNALITIES AND THE GROWTH OF TOURIST ACTIVITY

2.1. External economies and tourism

We have mentioned above that the tourist "product" is more precisely a set of goods and services. It is now useful to observe that for each economy of interest we can obtain or construct a vector of goods consumed or demanded by tourists in a certain period. This entails determining the goods and services acquired by tourists and thus determining their respective quantities. This vector is essential for evaluating the economic impact of tourist activity. For example, by using an input-output model or a computable general equilibrium model – based on economic accounting data from that particular economy – we could simulate the effects of an exogenous increase in tourist flows by means of multipliers, or of certain public measures regarding tourism (tax reductions, incentives, etc.).

Though important, the models based on the vector of goods and services acquired by tourists are not able to bring to light a very important problem: that of the services that are not exchanged on the market; in other words, those involving public goods and externalities. The vector of goods and services that each individual tourist makes use of includes non-purchasable services as well. The latter are collectively-enjoyed goods; that is, they are non rival in consumption, and they often represent the essential component of the tourist product, or in any event one of its most important components.

The sun, the sea, the mountains, the panoramas, the national heritage sights are extremely important to tourist demand. For example, we can see in Table 1.4 that national heritage sights in Italy account in a certain way for 29% of tourist demand, while the seaside brings in an additional 25%.

We can discuss whether these collective goods are directly enjoyed by consumers as we have stated, or if instead they are part of the productive factors of the tourist product: in any event, this is merely a nominalistic question on which it is not worth focussing attention. We can immediately see that some of these services (whether or not they can be considered as intermediate or final consumption goods), though collectively enjoyed, can also be excludable. In other words, we can exclude from the enjoyment of the goods those who are not willing to pay a price.

With regard to the possibility of exclusion, economic theory tells us that this feature is not important in terms of the conditions of efficiency; that is, for each pair of goods there is always equality between the individual sums of the marginal rates of substitution and the price ratios. But in reality these conditions, which imply individual prices, do not appear to be of help, except to highlight a case of market failure. On the other hand, the possibility of exclusion creates a situation, as we have just said, whereby a price can be paid by those who actively demand a particular service. If a specific tourist service could be supplied at zero cost (for example, gazing out over an inspiring panorama from a certain angle), then the price appears to represent a true Ricardian rent. However, very often, and especially for national heritage items, the tourist service does not coincide with the national heritage item itself, but is instead a specific product that is obtained "by using" the national heritage item together with other productive factors. Consider, for example, a museum, whose service involves capital and labour in addition to the exhibited paintings. The possibility of exclusion is important in this case, since it allows for the introduction of a price that in turn can serve to pay those productive factors.

Moreover, for many categories of national heritage items excludability plays an important role with respect to the conservation of this item itself, beyond the payment to the other productive factors that take part in the production of the tourist service. A museum is in fact a structure whose aim is not only to benefit from paintings but also to preserve them, and preservation is mainly based on the possibility of exclusion. In this case the fact that the entrance fee to the museum generally does not cover the management costs depends on the fact that it is recognized that a museum supplies a useful service not only to the visitors (tourists and residents) but also, due to its own existence, to the local collectivity, independently of the fact that it is visited. This is a case of a "mixed good" whose cost must be divided between the visitors and the local collectivity; from this point of view significant public measures regarding the production of tourist services are justified.

Another important feature of many collective tourist services is that there is often congestion; in other words, above a certain number of participants they become rival in consumption, at least partially. A beach, a museum, a road, or a park are examples of excludable services that give rise to congestion if there is excess demand with respect to the capacity for satisfying consumption. We know that congestion can be viewed as a specific negative externality, so that in such cases the price can also serve to internalize the marginal damage from congestion, with the ultimate effect of reducing the demand to the level of the efficient equilibrium.

Apart from the case of congested services, externalities play an important role in the production of tourist services; it is thus appropriate to briefly discuss this phenomenon. As we know, there is a consumption (production) externality when in an individual's utility function (production function) there appears a variable determined by another individual. Externalities cause the market to fail, in the sense that it is not efficient, and thus prices no longer represent correct signals, making public intervention necessary to internalize the externality. In other words, considering in particular external diseconomies, the production or consumption of tourist services can imply the joint production of public goods (final or intermediate) having a negative utility. In this case, independent of the legal system for the attribution of rights, the prices of tourist services will not correctly incorporate all the relative costs, and the resources will be allocated in a distorted fashion. For example, jet engine noise in the vicinity of an airport represents a negative externality linked to tourism which, in a system that protects the rights of polluters, does not influence air fares, and thus the price of tourist services.

Tourist services are particularly tied to the environment; not only – as we have mentioned above – is the territory, a primary component of tourism, but, as we have just stated, in the vector of tourist goods there are collective services offered free of charge, which in large part are closely tied to environmental aspects or features. It thus follows that environmental diseconomies can easily (or inevitably, to be more correct) result from tourist activity, whose consequences are more important the greater the production of tourist services is in a particular territory.

Moreover, it is also true that tourist activity can be subject to external economies or diseconomies caused by the production or consumption of goods aimed at residents. The simplest example of this is a factory that pollutes a valley dedicated to tourism. With respect to tourist activities, externalities can thus be active (caused by tourism) or passive, if caused by other activities that have an effect on tourism. Since the existence of externalities causes allocative or distributive distortions – there arises the problem of eliminating these distortions, in other words internalizing (or controlling) the externalities, which can only be carried out by an "external" authority. The external authority that is able to control the externalities is usually a public entity, either a central or local government body. This depends in large part on the characteristics of the specific tourist externalities. However, in general the public body is the only one capable of controlling the externalities, due to the coersive powers of the public authorities. Public intervention is justified from the economic point of view when it reduces the distortions caused by uncontrolled externalities.

In particular as regards tourist externalities, active or passive, we can conclude that their control is essential not only for preserving the environment, but also for the continuation of tourist activity itself. If a territory attracts tourists due to its environment, it is obvious that heavy deterioration of the environment would keep tourists away and cause the production of tourist services to cease.

The externalities connected with tourist services can also be positive; in other words there can be external economies. These must not be confused with the *economic advantages* from tourist development, which take the form of a growth in national income and employment, in addition to the inflow of foreign currency from the international component. For example, if tourist development leads to new hospital construction, this will certainly benefit the native population as well, and thus represents a positive externality. At the same time, the *economic* external diseconomies from tourism must not be confused with economic disadvantages. For example, the inflation that may result is a negative consequence of tourism, but it is not an externality, while the increase in car traffic due to tourists represents a negative externality of tourist activity.

The concept of externalities can also be extended beyond the strictly economic context, since tourism is not only an important economic phenomenon but also brings with it important social and cultural features. By definition, these aspects are not easily appreciated or valued from the economic point of view, but they are nonetheless important, since the economic component, though important, represents only one aspect of society. For example, tourism brings various cultures into contact, and this is certainly a positive fact in terms of cultural enrichment. Nevertheless, mass tourism can lead to the loss of cultural identity or local traditions, and this is decidedly a negative aspect. It can also lead to the formation of artificial entities, where the local folklore that attracts the tourist is tailor-made for the tourist, as in the case of the Bali syndrome[5]. Here it is more difficult to decide whether there is a positive or negative influence. At least in a broad sense we can say that there are external effects whose economic valuation nevertheless is not only difficult but also not very appropriate. These external effects cannot, however, be ignored by the public, since they can have quite important consequences that, at least within certain limits, the governments must try to keep under control by means of appropriate measures.

The latter basically involve measures aimed at creating advantages or disadvantages for tourist activity in general, or the individual aspects of such activity, and as such cannot substantially be distinguished from public interventions aimed at controlling the economic externalities. In fact, just as the economic externalities can be controlled by meta-economic means (regulations), external socio-cultural effects can be controlled by economic means. Public measures can control the externalities or the external socio-cultural effects, though it is not possible to arrive at the cause that inspired the intervention by knowing only the intervention itself.

2.2. The conflict between the control of externalities and growth

In fact, there is little concern regarding the (positive) external economies affecting tourist activity or, conversely, those deriving from tourism and directed toward the economy as a whole. This is probably due to the fact that there are few such external economies and these have little importance from the economic point of view. On the contrary, more attention continues to be given to the external diseconomies, both active and passive.

As far as the latter are concerned, there exist important examples, such as the activity of the chemical industries in the Dead Sea (Israel) which have caused a rise in the sea level of 20 inches per year, greatly damaging the bustling tourist activity of the region[6]. But we can say that it is above all the negative externalities which attract attention; that is, the problem of the external diseconomies from tourist activity is increasingly felt to be urgent. In particular, the negative environmental externalities have taken on prominent interest; in other words, the negative repercussions of tourism on the physical, social, and cultural environment.

The fact is that economic systems today make considerable use of the environment, above all as a dumping ground for the by-products with negative utility. Remember that the production of tourist services represents an important part of overall production, and thus contributes proportionally to this discharge. There are two salient features of this industrial use of the environment: on the one hand, the nature of the goods "discharged" into the environment no longer permits biological recycling by the environment (plastic material, etc.). Secondly, the quantity of this waste has greatly increased, reaching proportions that make natural recycling impossible. Mass production and modern technology have increased the production of waste to the point that the environment can no longer handle it through natural recycling. In all ways the environmental damage caused by tourism is not limited to waste as such: we need only consider the cementing of coasts or the changes brought about in ecosystems to realize that tourist development now represents a serious threat to the overall environment.

The large increase in the industrial economic process has been accompanied by the rise and development of mass tourism, which in turn is linked to the growth in national income, the increase in productivity, and thus the increase in free-time. We have seen in Table 1.1 that the international tourist movement is still developing at a fast pace: over 600 million tourists world-wide, with an annual growth rate between 3% and 5%. We also know that most tourism is domestic, which is the type destined to develop most from an increase in national income. As far as the future is concerned, much depends on the scenario we can envision: a reasonable assumption is that both developing and developed countries will maintain the growth rate trends for national income seen over the last decade. In this case even tourist movements should increase at more or less the same rate, and the consequences for the environment will be increasingly serious; public intervention thus must become increasingly important as well.

Such public intervention takes place according to the traditional means of controlling the externalities; that is, regulation, taxes, and subsidies. Moreover, given the breadth of the intervention itself, it is essential to dedicate important efforts to the governmental management, control and coordination of all tourist activity. In other words, tourist activity should not only be controlled but placed in a planning context that is coordinated on the one hand by monitoring the growth of national income, and on the other by environmental protection actions.

From this perspective a problem arises from the fact that the means used by governmental bodies to control tourist externalities (even in different ways depending on the specific objectives pursued) tend to contain or discourage tourism, or at least its specific aspects. This means that public measures aimed at controlling the externalities could imply strong interference with public measures that seek to develop and sustain tourism; these measures are traditionally important since this sector has become, and continues to be, increasingly important for the growth of national income.

In fact, in all countries concerned the entire traditional structure for public intervention in the tourist sector is directed toward tourist promotion, both at the central and local government levels. We know in fact that tourist marketing is essential to launch and maintain the image, especially with the growth in competition. In these conditions governmental promotional measures are particularly appropriate, even if it is true that large tourist multinationals can launch new tourist initiatives even without public intervention. Public intervention is appropriate once tourist activity is determined to be "advantageous" for the collectivity; promotional activity favours all those involved in the sector, including those who do not contribute any money (free-rider behaviour). In this case "private" promotion would be below the optimal quantity; moreover the advantage would not be limited to tourist operators alone but would involve the entire collectivity in the form of an increase in national income, employment, and foreign currency earnings. This represents an additional reason for public promotional measures.

This situation implies a more or less latent conflict between the traditional measures aimed at promoting tourism and the more recent measures aimed at controlling the tourist externalities, since, as we have said, the former tend to develop and the latter to contain tourist activity. This conflict can be resolved only by a careful global planning effort involving all aspects of tourist activity.

Even the situation in Italy presents a regulatory context that is mainly oriented more toward the promotion of tourism than toward its planning and the control of its effects on the environment. We can add to the tourist development-externality control conflict another source of conflict, that which is implicit in the State-Region dualism with regard to tourist measures. As we know, in Italy the latter came under the authority of the regions under article 117 of the Constitution, which was subsequently strengthened by the outcome of the 1993 referendum that did away with the ministry of tourism. Nevertheless, the state maintains an important role in this area, with a substantial continuity that began with the outline law in 1983 and has continued until the recent Bassanini law; this law, while not explicitly mentioning tourism, regulates this sector together with the other matters included in article 117 of the Constitution. It is possible that the reorganization owing to the carrying out of the Bassanini law will mean that in future the state-region conflict will be replaced by constructive dialogue. However, for now the situation is unsatisfactory: "there has not been any regional law of structural importance that has not been radically modified by a subsequent state law, thereby undoing the work of the regions" (F. Indovino Fabris, 1997, p. 820). Moreover, we must consider that in the past the regions themselves have tried to propose a sort of regional centralization for the provinces and comunes that is modelled on state centralization[7], which in turn implies conflicts and a lack of operational incisiveness.

At present, the regulations spelled out in the law 203/1995 give the state authority with regard to general decisions by means of the Government and the presidency of the Council of Ministers, while the operative functions are centered in the Department of Tourism, which is under the presidency of the

Council of Ministers, and the ENIT, which has promotional responsibilities. Coordination with the regions is guaranteed through the permanent conference of the state, regions and autonomous provinces, as modified by the Bassanini law. In fact, until now the regions have dealt with the legal aspects of the tourist profession, on the one hand, and with the development of the tourist sector on the other. As far as the tourist profession is concerned, the action of the regions has been limited by measures by the state and the European Union, and it does not appear that on the whole the region's role has been satisfactorily worked out. As far as growth is concerned (without considering the results obtained), since the efforts have been concentrated above all on the Tourist Promotion Agencies, nearly the entire focus has been on the promotional aspect. Thus the role of the regions has been a limited one with regard to the true needs, which concern the direction and coordination of tourist resources at the regional level.

Future development depends on how the Bassanini law is applied. On the one hand this law, while maintaining a strongly unitary structure from the constitutional point of view, decentralizes or delegates responsibilities to the territorial bodies according to the subsidiarity principle, while on the other it involves them in the decision-making process. The main problem is, identifying, creating, developing, and taking advantage of tourist resources, as well as maintaining these resources. However, in order to effectively achieve these objectives a coherent and concerted regional policy is needed that includes the lesser territorial bodies; a policy that, starting from the fundamental decisions, can be applied through a suitable means at the territorial level.

3. TOURIST ACTIVITIES AND THE SUSTAINABILITY OF ECONOMIC GROWTH

The intervention of the state is thus necessary to control the various negative externalities caused by tourist activities, even though this can seem to be in contrast to the growth needs of these activities and thus to economic growth in general. In reality the control of the negative externalities, especially environmental ones, has the aim of re-establishing the conditions of efficiency of the economic system and thus favouring the welfare of the collectivity.

In the tourist sector firms usually tend to have a short-term perspective; motivated by the objective of maximizing profits, they tend to favour an increase in tourism without taking account of the territory's ability to support such inflows of tourists, thereby often causing a loss in the attractive capacity of tourism. This leads to a true cycle for the tourist product; in fact, the growth of tourist activities is characterized by a growth phase that is first slow and then more rapid, with an income elasticity greater than one, and which subsequently becomes stable or even begins to decline due to congestion, environmental degradation and, in general, a decline in the quality of services (Butler, 1980).

This effect is especially seen with regard to tourism based on the existence of natural resources. With regard to these resources we witness a *snob-goods* effect, where the demand for a tourist good is not only a function of its price or the cost of a visit but also of the number of other tourists taking part (Leibenstein, 1950). In this case, given tourist aversion to traffic and decaying resources, tourists will be less willing to pay, since the quality of the resources, and in general that of the tourist locality, declines the greater the number of tourists there are. Thus, contrary to normal goods, where the consumer surplus increases as the price falls, for goods whose demand is also determined by snob appeal the consumer surplus falls as costs fall or, in any event, environmental decay and congestion decrease. It should nevertheless be pointed out that tourists are not always averse to congestion: due to a sort of bandwagon effect, the existence of overcrowded tourist goods can attract other tourists (Tisdell, 1991).

In fact, a decline in the number of tourists can be linked to causes other than environmental decay. For example, again in terms of product-cycle theory, this can be due to competition from other localities that offer the same type of attractive services.

The state must nevertheless intervene to impose a long-run perspective that aims to ensure growth possibilities for the sector. This intervention is even more important when environmental resources – natural, climatic, panoramic, or consisting of the artistic and cultural heritage of the area – play a significant role in tourist activities.

Thus, in defining the various forms of intervention we must keep in mind the features of tourist goods and the various forms of tourism associated with them. Tourist goods can in fact be reproduceable – for example, hotels, campsites, transportation, etc. – and non-reproduceable, when they depend essentially on the availability of environmental goods. The latter, which have mainly an inelastic demand, represent a very important component of the overall demand for tourist goods and services and, as we have mentioned, are characterized by the limitation of the available stock and by the fact that, even when they can self-produce (fishing resources, the assimilitative capacity of the environment), this is dependent on the preservation of a part of the stock.

The public intervention that is necessary for both categories of goods and services has more than a few problems, especially as regards non-reproduceable environmental goods, since these have no market price and are used intensively in a way that can cause their irreparable destruction. In this case we must identify and permit a level of resource use that is compatible with the conservation of the available stock so as to ensure the satisfaction of future consumption and thus not compromise the potential flow of future income from its use for tourism. In other words, the management of these resources must also guarantee for future generations the chance to enjoy them and ensure that the activities which are based on their use can continue and contribute to a sustainable economic growth.

The problem seems to boil down to determining the optimal level of resource exploitation, which is given by the maximization of the net income obtained from such exploitation. It is thus enough to evaluate the costs and benefits from exploiting the resources and to determine an appropriate discount rate that equalizes the flows of these costs and benefits. However, the solution to the problem is not easy since environmental resources have no market price; in any event, the problem of sustainable growth cannot be circumscribed by determining from a purely economic point of view the optimal level of exploitation of a stock of resources.

In fact, the concept of sustainability, which already came to the attention of public opinion in 1987 thanks to the UN World Commission on the Environment and Development (Bruntland Commission), does not only concern the use of natural resources. In defining sustainable growth as "a process of change in which the exploitation of resources, the direction of investments, the orientation of technological development and institutional changes are made consistent with future as well as present needs" the Bruntland Commission was referring to a vaster process of development that includes, in addition to economic aspects, institutional and social ones as well (WCED, 1987, p. 43). In particular, the definition of sustainability revolves around two concepts: "equity" and "compensation".

As far as the concept of equity is concerned, sustainable growth requires there to be equal opportunities for future generations to satisfy their needs (intergenerational equity), but also equal access to resources for all individuals no matter where they live, including therefore the most disadvantaged individuals (infragenerational equity). Generally the recognition that future generations must have the same opportunities as present ones is resolved by the use of a series of intertemporal obligations or constraints that can take various forms: consumption or utility that is non-decreasing over time; non-decreasing total capital (wealth) stock; non-decreasing natural (environment) capital stock, etc. (Pezzey, 1989).

A tourist development strategy could be sustainable if it permits resources to generate in the future at least as much income as they can produce today (Pigliaru, 1996). In this case, and going back to Hicks' definition (1946), income would be measured as that flow of goods and services that the economy could produce without reducing its own productive capacity. This notion of sustainable income captures the idea of a constant capital stock, both physical and environmental, which is at the center of most discussions on sustainable growth. We can just as well define sustainability of tourist activity so that "the demand of increasing numbers of tourists is satisfied in a manner which continues to attract them whilst meeting the needs of the host population with improved standards of living, yet safeguarding the destination environment and cultural heritage" (Nijkamp, Verdonkschot, 1995, p. 127).

According to these definitions, sustainable growth could also be compatible with very intense exploitation of environmental resources, and with large reductions in their quality over time. But beyond the definitions, which are considerably important for determining the conditions that regulate the sustainability of economic activities, what must be pointed out is the underlying logic: future generations, as well as individuals from the same generation, should have the same opportunities to satisfy their needs. If for some reason they do not have these opportunities, there must be compensation, and this represents the second concept around which the definition of sustainable growth revolves. It is the way that this compensation should take place that is at the center of the debate on sustainable growth, and that should influence the intervention strategies of public and private sector workers in the field of tourism.

4. ENVIRONMENTAL RESOURCES AND THE SUSTAINABILITY OF TOURIST ACTIVITIES

In defining these forms of compensation the distinction is usually made between *weak and strong sustainability*. For the former concept it is necessary for total capital, manufactured and natural capital, not to diminish over time, while for strong sustainability the emphasis is on the preservation of the natural capital stock and requires that this not be smaller than that enjoyed by the present generation.

The first definition requires there to be perfect substitutability between natural and manufactured capital; thus little or no importance is given as to whether or not the environmental goods diminish, as long as an equivalent amount of man-made capital – that is, reproduceable tourist goods and services – can replace them. This idea goes back to Solow (1986), according to whom, under the assumption of perfect substitutability among the various capital components, the present generation does not owe future generations a particular share of capital but, instead, *access to a certain standard of living or level of consumption,* independent of the form in which this is conferred. Society as a whole can improve its condition by exploiting natural resources, under the condition that it uses the proceeds of this to create a stock of other goods. The assumption implicit in this definition is that the man-made capital does not diminish as it is passed from one generation to another; the stock of natural goods can diminish as long as the growth of man-made capital compensates for this reduction.

Technological progress will allow substitutability between the two forms of capital to be realized. In this case environmental concerns, though present in development policies, do not in any way represent a constraint for the pursuit of sustainable growth. On the basis of this first approach there would thus also be the possibility of guaranteeing the growth of tourist activities by giving up part of the stock of environmental goods that represent part of the supply of these activities.

In fact, with regard to tourist activities that are primarily based on the endowment of environmental resources, pursuing a growth objective based on the concept of weak sustainability would create a number of problems, due to the unevenness of technological progress. The latter operates more in favour of reproduceable tourist services than of environmental goods. Technological progress can increase the total stock of capital in the economy and thus the capacity of the economy to produce goods. Nevertheless, it is *far less capable of enhancing the supply of wilderness* (Krutilla, 1967). The unevenness of technological progress leads to the increasing production of man-made goods but fewer natural goods.

Technological progress certainly contributes to increasing efficiency in the use of resources. Nevertheless, as far as environmental resources are concerned, there are problems of irreversibility that condition their choice. Contrary to the case for reproduceable goods, technological progress is not able to increase the supply of natural goods. The supply of environmental goods can be considered to be largely inelastic and in general non-expandable with regard to technological progress. It thus follows that over time the availability of reproduceable goods will increase relatively more than the availability of environmental goods and, as far as relative prices may change to reflect a greater desire for environmental goods, there will be no possibility of moving in the desired direction. Future generations will thus find themselves with a smaller stock of environmental goods than that available to preceding generations, thus compromising the possibility of sustainable growth.

The problem is illustrated in Figure 1.1, where the environmental goods are along the vertical axis and the reproduceable ones along the horizontal axis. The curve AB represents the transformation possibilities of the economic system in period I, and since we assume constant returns in the economic system the curve is a straight line. The curves i, i', i", i''', etc. represent indifference curves that describe a social welfare function (Magnani, 1974). Equilibrium is at point E, where the transformation curve AB is tangent to the highest social welfare curve. At this point all the optimality conditions are satisfied; in particular, the price consumers are willing to pay is equal to the slope of the transformation curve, and thus to marginal cost.

Let us assume that over time technological progress, which concerns only reproduceable goods, acts to shift the transformation curve upward, rotating around point A in the positions AD, AL, and AM, respectively, and so on. In other words, this determines an increase in the prices of environmental goods with respect to other goods, which means that the former (the reproduceable ones) are relatively less scarce than the environmental ones. Over time and with an improvement in technology the new equilibrium points will be determined by the tangency points of the transformation curves with the corresponding indifference curves that can be attained at various points in time.

The loci of these points, which is described by the curve ß, represents the optimal expansion path along which all the optimality conditions are satisfied. The problem is that the economic system is not able to follow that path. Along the section AF environmental goods should be transformed into reproduceable ones, but from point F onwards it would be necessary to transform reproduceable goods into environmental ones, due to the greater overall scarcity, greater income, and more advanced technology; however, such a transformation is impossible or, in any event, too costly.

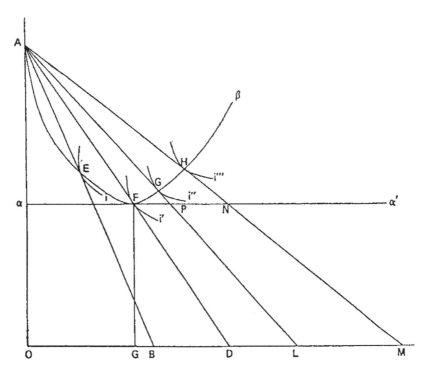

Figure. 1.1. Technical progress and environmental resources

The increasing section of the growth path Fß implies an increase in the desirability of both environmental and reproduceable goods. This is due to the fact that an increase in income tends to lead to more consumer demand for environmental goods, given that these are considered to be superior goods. Beginning from point F the economy will be forced to move along a path that is different from the optimal one, Fα for example. Economic theory has shown that this result is highly probable due to the characteristics of public and environmental goods and, in general, to the negative externalities from production and consumption activities, which lead to an overuse of environmental resources (Magnani, 1974). In addition to these externalities there are those that concern the uncertainty in the direction of technological evolution, or consumption externalities that involve the position of the indifference curves.

Given this situation, and since it is unlikely the market will be able to remedy the scarcity of environmental goods in the absence of constraints or, in any event, interventions to conserve environmental resources, tourist activities based on these resources could be compromised. It is thus necessary to define the use level that is compatible with the preservation of tourist activities, and the second sustainability scenario, commonly defined as strong, goes in this direction. Several environmental goods, such as landscape, space, and tranquillity are essential for human welfare; not being easily substitutable, according to the rule of strong sustainability they must be protected.

This way toward growth seems to be more plausible if we consider that, with an increase in income and greater scarcity of environmental goods, tourist preferences shift toward high-quality, non-crowded goods based on natural resources (McConnell, 1966). The possibility of maintaining a growth or welfare potential for future generations requires the application of management principles which are specific to each of the wealth components that must be transferred.

The maintenance of the stock of environmental goods thus becomes one of the conditions for sustainable growth. The extent to which this condition is less pressing depends, in addition to the role of technological progress in reducing the need for environmental goods as inputs for an improvement in the standard of living, on the effects of the increase in tourism on the deterioration of the stock of natural capital, and, lastly, on the preferences of present and future consumers.

5. THE TOURIST CAPACITY LOAD OF THE TERRITORY AND RELATIONAL GOODS

The trade-off between environmental resources and reproduceable goods pointed out above is thus not only a technical problem but, in the final analysis, depends on the preferences of tourists and the changes in these preferences over time. Moreover, this trade-off is limited by the capacity load of the territory, which is the maximum number of visitors the area can hold without an unacceptable alteration in the environment (natural and cultural) and an intolerable deterioration in the quality of the tourist experience such as to reduce the tourist flow.

In fact, this quality should be understood not only from an environmental point of view but an economic one as well: it should include the qualitative features of the environment as well as the economic activities of the territory. From the environmental point of view this involves the functions of the environment and the endowment of environmental resources, while from the economic point of view it refers to the sustainability of the area for carrying out tourist activities without compromising essential activities for the local community. Determining capacity load thus requires coming up with quality and quantity indicators that represent the stock of environmental resources and its variation, the territory's economic activities, the relationship between these activities and tourist ones, and thus the role of the actors operating in the territory.

It should be pointed out that the trade-off between environmental goods and reproduceable tourist goods can end due to the existence of a quality floor in the territory from an environmental and social point of view, and, in general, to the capacity load compatible with the territory's tourist management. In other words, there are choices that make economic activities incompatible with a balanced management of the territory, since the production of goods is obtained in a situation where the quality level of the environment is below the minimum that is compatible with the growth possibilities in the area.

If we do not take account of the territory's carrying capacity, the solution that calls for the substitutability of environmental goods with reproduceable ones could cause a loss in the comparative advantages of tourist areas. This outcome can obviously be avoided in part, especially if technology and the dematerialization process that we have seen in the last decade with respect to economic activities is able to reduce the impact on the environment (Gerelli, 1995). The extensive spread of new processes, product or organizational technologies, and especially information and high-efficiency material technologies, has in fact contributed to substantially reducing pressures on the environment. In this case more advanced equilibriums could be achieved that correspond to a larger quantity of environmental goods and, above all, tourist services, if the production possibility curve shifts upward due to technological progress; as we have already said, this can be achieved through the use of clean technologies, or in any event with organizational systems for the supply of tourist goods that tend to reduce the impact on the environment. Nevertheless, we must still pay appropriate attention to the carrying capacity of the area in question.

The various decisions regarding the growth of tourist activities must thus be made in the context of an overall plan for the use of the territory's material and human resources that has as its objective not only the exploitation of the economic potential of tourism but also the definition of compatibility among the economic activities as a whole and the area in question. Of great importance in this context are the communications infrastructures and, in general, the information system that is capable of achieving, among tourist operators, a horizontal relational system among the local actors for the sharing of common objectives, as well as a vertical system with the other territorial systems.

The production of relational goods is important also in light of what was said above about the trade-off between the utilization (at times irreversible) of environmental resources, the advantageous economic use of resources, and tourist capacity. In fact, the problem concerns short-term benefits for entrepreneurs and consumers and the welfare of the collectivity from a more long-term perspective. In this sense the environmental consequences from quantitative and qualitative changes in the characteristics of supply and demand should represent crucial elements for evaluating tourist policy.

The creation of a system of relations among the various operators can be achieved on the basis of an information system capable of permitting the individual actors to carry out a systematic measuring of the main costs and benefits from initiatives taking place in the territory, thus allowing the operators to assess the economic advantages of these initiatives and to undertake investments in the territory. With regard to these needs we should note that in general there is a serious lack of data, statistics, and base indicators, as well as other quantitative and qualitative information necessary for evaluating the environmental conditions and their tendencies, the endowment of economic resources, and the operational methods of enterprises in the area.

The same translation of sustainability goals into actions requires the determination and measurement of sustainability indicators. The complex and still debated aspect is over what kind of sustainability indicators should be used to design economic policies for achieving both environmental conservation and societal goals. With regard to this last problem, we have already seen the importance of the definition of carrying capacity, and how this depends on the size of the supporting territory, the endowment and quality of resources, the type of activity located there, and the available technologies. The information that is appropriate for defining the carrying capacity can thus be organized as follows:

- natural resource stock indicators, such as changes in forest and agricultural areas, forest stocks, land and population area, and the amount of land in its natural state; the amount of land in residential, industrial, commercial and recreational use; percentage of protected areas, and ratio of urbanized to non-urbanized land; distribution and type of vegetation, etc.;

- economic indicators of natural resource scarcity, such as the price of urban land;
- environmental quality indicators: air quality, quality of surface water, congestion, level of noise, etc.;
- quality of life indicators, such as population density, the state of the protection of cultural and archeological sites, etc.

6. THE CONDITIONS FOR SUSTAINABLE ECONOMIC GROWTH

The sustainability of economic growth is thus a concept that does not only concern natural resources but includes as well the various elements that constitute the economic and social system as a whole; thus the definition of a model of sustainability requires us to consider both economic activities as a whole as well as the social and, of course, environmental systems. The problem involves considering these systems with regard to their dynamic interactions and not simply as a sum of their various aspects.

Looking at the problem from an operational point of view, and basing the definition of sustainable growth on the constancy of the parameters, in particular on natural capital alone, could create some difficulties from a long-term perspective; in fact, economic growth has a positive cost. Over time the interrelations between economic, social and natural environmental systems produce a multitude of transformations. The taking hold of tourism represents an additional factor with respect to the preceding situation that tends to change the pre-existing economic, social and territorial equilibrium, causing in some cases problems for the environment and for other activities that have previously taken root. The change is greater the more important the role of tourist activity in the local economy. In particular, territorial tensions and the deterioration of employed resources are closely dependent on the spatial and temporal distribution of resources and on the degree of specialization of tourist activities.

It is thus difficult to imagine a concept of sustainability based on the strict observance of the rule concerning the constancy of natural capital. We must in fact keep in mind that the physical trade-off between natural and artificial resources could bring about more efficient territorial processes; thus, from this perspective, it does not seem right to include among the arguments for sustainability the absence of a trade-off between these systems, as long as those constraints represented by the carrying capacity of the territory are respected, together with the preferences of consumers.

The non-substitutability between natural and artificial capital, which seems to be the most appropriate assumption, can, from a global point of view, turn out not to necessarily be the best solution at the local level. The sustainability of economic growth, though focussing attention on global competitiveness, should be defined and carried out taking account of the specific features of the territory and the environment. This implies that we must rightly consider several models of sustainability. In order to concretely pursue the sustainability of growth it must be territorially located, since carrying capacities vary, as well as the potential of each country. "History has shown there aren't resource-rich and resource-poor countries, but countries whose economy and knowledge is suited to its resources and those with a growth model that is extraneous to their territorial and environmental situations. Resources as well as sustainability are created starting from knowledge of one's environment and basing the choices for growth on this" (Bresso, 1993, p. 102).

In operative terms, achieving sustainable growth, which represents a future objective, must thus take into account the integration and synergies within the territorial system, which is composed of several local subsystems (economic, social, environmental), and of the relation between the latter and the global system. Growth capacity thus translates into the different capacity of local systems to combine and organize economic opportunities, resources and actors in order to position themselves most advantageously as regards territorial reorganization processes.

The development of tourism, while it depends on local environmental and socio-economic conditions, must measure itself against a wider territorial context, above all regarding the globalization of economic systems. Competition thus does not occur between firms or groups of firms, but between organized territorial systems. From this perspective it can depend on three sets of factors:

- the international opening of territorial systems with the related capacity of facilitating worker and capital mobility and lowering transaction costs with other systems;
- a high endowment of resources (environmental, economic, social, and cultural);
- the capacity to organize and take advantage of territorial identity.

Inevitably the international opening of territorial systems ends up conditioning the structure of enterprises and thus that of the market. The enterprises, which are searching for greater profitability, are called on to engage in relations that are not necessarily limited to the local context, and to take measures to achieve both vertical and horizontal integration.

Environmental endowments represent the most important conditions for dealing with the present phase of economic transformation. Beyond and preliminary to business innovation is environmental innovation, in terms of a strategic resource enterprises can count on. Policies should thus be directed toward improving environmental resources and, more generally, specific local features, and not toward reproposing solutions which are valid for completely different situations.

In order to undertake such measures appropriate government structures are needed capable of creating the necessary synergies among the various actors operating in the territory. The main task consists, as mentioned above, in promoting and developing a system of relations. Promoting agreements among enterprises represents a response to the challenge of market integration, to the complexity of the relations of interdependency among firms, and to the processes of technical and organizational innovation. Of importance in this regard are those accords undertaken by small and middle-sized firms interested in maintaining a market presence the individual firm is not capable of maintaining.

However, the problem of agreements also applies to the public entities charged with triggering economic growth and advantageously positioning the local territorial system. These entities must create, at the various territorial levels, new social contracts; that is, a system of relations that creates a climate within which the economic actors can obtain the resources needed to compete.

In concrete terms, as will be more clearly illustrated in the following section, the application of this relational strategy can be achieved by means of three types of intervention: that aimed at creating information systems; that aimed at improving the capital, and more generally at promoting technological innovation; and, finally, that aimed at conserving and exploiting the physical and natural environment. These measures must be translated into specific actions whose definition and attribution must be determined on the basis of clear guiding principles that ensure that these actions are carried out efficiently and effectively and respect territorial constraints. These principles are:

- *Precautional principle* – this principle explicitly recognizes the problem of uncertainty and tries to avoid the irreversible damage that economic activities can do to the environment through the imposition of a margin of security in the definition of economic policies.

- *Specificity principle* – the specificity principle involves the temporal and spatial validity of the various actions. What is valid at a certain time and in a certain place is not necessarily so in another moment and place.

- *Subsidiarity principle* – the subsidiarity principle, which is mentioned in the Fifth Program of action of the EC (1992), calls for the search for the most efficient level of government in terms of the objectives to pursue. This principle has both vertical and horizontal validity. Vertically, among the different levels of government it recommends transferring responsibility and resources to government levels close to citizens, leaving to the higher levels the tasks which the lower ones cannot adequately deal with. Horizontally, in public-private sector relations, this principle takes advantage of the autonomy of civil society, not only furthering decentralization but also lightening the load of the state and public administration. In this case the sense of the principle is in its assigning to public action the task of providing only those services and meeting those needs that cannot be effectively guaranteed by the market or by civil society.

- *Principle of co-responsibility* – the launching of a sustainable growth plan, as for that matter any growth plan, is necessarily tied to the amount of consensus it can obtain. Thus, the success of actions aimed at achieving sustainable growth depends to a large extent on the definition of objectives, the actions taken by the territory's actors, and shared support for these actions. It is thus necessary to define a global strategy that defines and coordinates the actions of the economic and social factions.

7. THE NEED FOR AN INTEGRATED POLICY

The management of tourist activities thus needs to be supported by various measures that, in economic terms, are able to influence both the demand and supply side of tourist activities. From this perspective the management of tourist activities does not depart from the management criteria of any other productive activity which seeks to maximize profits.

As far as demand is concerned, measures must aim at expanding tourist consumption, distributing this over time and space. With regard to the supply-side, measures must aim at creating the conditions for facing international competition. In particular, the supply of a product must, as much as possible, have features that coincide with the products demanded and, in addition, the supply of activities must be able to accommodate the modifications that may come from the demand side.

This presents particular difficulties, given that tourist production presents certain rigidities. We have seen that certain productive factors are not reproduceable (environmental and artistic resources) and that certain activities cannot be stocked, meaning the consumption of the product must occur where it is produced. Thus supply cannot satisfactorily adapt itself to demand, which from time to time will make it impossible to fully satisfy demand or will lead to excess capacity. Again with regard to the problem of supply-side rigidity, we must point out that, within certain limits, investments are irreversible in nature, in some cases representing true sunk costs. Not having alternative uses or, in any event, not being able to be quickly reversed, such investments can create considerable problems when demand is falling.

Measures to boost demand should be shaped to allow firms to improve the profitability of their production by reducing costs and possibly permitting an expansion of output if the reduction in costs leads to a fall in prices.

The reduction in costs can be achieved by taking advantage of economies of scale that derive from the integrated operating methods of firms in the sector. The product can no longer be considered as a set of individual services but as a system with vertical integration among the productive activities that play a supply role for the production of final consumption goods, as well as a system with horizontal integration among tourist operators in the territory. There will be problems for small and medium-sized firms that are not always able to exploit globalization processes underway in the economy. Those firms unable to deal with these processes through vertical integration strategies and thereby maintain their price competitiveness must adopt a strategy involving product quality and the relational links among the territory's enterprises.

The need to improve qualitative services also derives from the need to face competition; thus these services should be produced according to qualitative standards that are at least not below those of competing countries. This assumes a more or less constant flow of large-scale investments and the possibility of financing these. Favouring the development process and the quality of tourist activities   requires finding adequate public and private financing. In fact, the massive investments required by tourist production make firms in this sector financially vulnerable, and thus the difficulties in finding the financial means needed to undertake these investments can represent a considerable comparative disadvantage if the financial markets are not adequately developed.

The quality process should include the entire territory and those activities involved in tourism, with the aim of exploiting the territorial identity, given that this represents the most important selling point with respect to other tourist destinations and is the factor that makes tourism a sustainable activity. Taking advantage of territorial identity plays an important role in dealing with competition from other areas.   The availability of non-transferable resources makes tourist exploitation possible due to the relative uniqueness of such resources, which permits monopoly or near-monopoly situations to develop. We must nevertheless emphasize that even the endowment of tourist resources does not represent a given in the context of international tourist flows: changes in consumer preferences can lead to changes in tourist flows.

Thus the measures commonly proposed to relaunch tourist activities are traditional in nature. With regard to these measures two particular types of policy options previously touched on should be mentioned: those that aim at creating an information system and motivating technological innovation and those intended to preserve and take advantage of the environment.

7.1. Technology and the quality of services

Providing an incentive for technological innovation should aim at reducing the costs of basic services (transport, reception capacity etc.) and distribution costs, and at improving the quality of the goods and services supplied, including environmental ones.   This can involve both an organizational and institutional process aimed at creating the necessary coordination among public and private operators and at providing a systematic approach to the measures, without which there would be no integrated operating method, which remains one of the fundamental assumptions for the growth of tourism. These incentives must also be aimed at offering products that differ from those offered by competing markets, and with better quality at a lower price.

The economic analysis of technological innovation processes has brought to light how the market, due to the unique nature of such goods (indivisibility and inappropriability), is not able to divide up in an optimal way the resources directed at the production of innovation (Arrow 1962, Nordhaus 1967). There is no reason to believe that the technological progress that allows for substitutability among the various forms of capital will be automatically and spontaneously introduced into the economy, or in any event that it will be optimal. The public sector must thus intervene to assure an optimal quantity of innovation.

In other words, firms must be motivated to invest in technological transformations that can reduce the environmental impact of their own activities. Specific investments are needed in order for there to be a continuous reduction in the environmental impact from production and consumption processes; thus the choices by firms must guarantee that an appropriate share of their overall investments go toward reducing environmental impact, thereby giving rise to less polluting production processes and final products that put less pressure on the environment. In fact, the growth of tourism does not necessarily mean a growth in its material production; it can equally mean a growth in the quality of production and consumption that translates into a growth in the economic value of production. From a static point of view this feature of economic growth is determined by the increase in the weight of advanced services and non-material production in the value of national income, and by a reduction in material production. This transformation is determined in large part by innovation, which, we have seen, is the basis for the dematerialization process in the economy.

7.2. Environment and territory

From the above considerations it thus emerges that the task of the public sector consists in providing incentives for conserving and exploiting the natural environment, and that this objective assumes the difficult task of determining the tolerance limits of the ecosystem and the related carrying capacity of the various areas. In order to do this it is necessary to define indicators that translate environmental constraints into operative terms and to insert these in a planning system that can coordinate the various measures that are necessary from time to time.

Territorial planning represents the means by which the various interventions in the territory have usually been coordinated. Above all with regard to the use of environmental resources when there exist problems of irreversibility, it is appropriate to turn to environmental constraints such as quality standards or quotas for the use of resources, and in any case to undertake any action that is compatible with the precautional principle mentioned above. Administrative instruments such as command and control are nevertheless critical due to their intrinsic rigidity, on account of which the use of means that permit the economic actors to operate with more freedom – by using, for example, economic instruments – are usually recommended. The latter refer to market incentives which, better than any other instrument, can give a price to those resources, such as environmental ones, that are not transacted on the market.

Public measures thus consist in correcting the markets by determining at central level the value to give to environmental resources and guaranteeing that that value is reflected in the prices of goods and services. Giving a price to those resources that do not have one or correcting inefficient prices represents one of the measures to take if the objective of sustainable growth is to be pursued.

Among the instruments available to the public sector, the economic ones – duties, taxes, tariffs or tax rebates – get across in a more immediate way the idea of a social price for the use of environmental resources and are preferred, at least in theory, due to the comparative advantages with respect to other instruments. These are instruments that are already used by the public sector and that take various forms: per capita tourist taxes imposed through hotels and aimed at collecting local revenue for services which are difficult to impose on the basis of use; local government taxes based on the value of property; charges for services according to use.

Nevertheless, it is a widely held view that public authorities do not adequately exploit the potential of such instruments and thus, among other things, miss out on increasing their tax revenue. It is also for this reason that, with regard to controlling environmental tax measures, we must be well aware of the possibilities presented by the tax autonomy available to local authorities, especially in light of the important role granted to them by the rules and regulations regarding environment protection.

There is great interest in the possibility of creating new markets for allocating environmental resources through the use of certificates that can take various forms: emission certificates, resource use quotas, construction rights, etc. According to Dales' idea (1968), the task of the public sector is to fix environmental quality or quantity standards relative to the use of certain resources that are to be protected, and subsequently give out a number of certificates allowing the bearer to use the resource that corresponds to the maximum quantity compatible with the previously set environmental laws. Each holder of these certificates would thus have the right to use an amount of environmental resources equivalent to the number of certificates possessed.

Certificates can be exchanged among resource users at a price determined by supply and demand; where there is perfect competition in the market, this price should settle at the marginal cost for the use of the resource.

The measures we have referred to should, if applied correctly, be equivalent. In reality they differ with regard to the costs of information, control, and management. Taking account of these costs, as well as other parameters, such as their degree of flexibility or, more generally, the incentives they can provide to the economic actors, economic measures are usually preferred. In fact, the latter tend to simulate the market, and through the use of prices (duties, taxes, tariffs, etc.) individuals are given incentives to behave in a way that is compatible with the objectives of the public sector.

We must nevertheless observe that, as regards the allocation of unique resources – which involves solving threshold problems – the use of economic instruments must be coordinated with administrative ones. While the command and control-type approach is indispensable for defining the minimum level of environmental protection (minimum environmental quality standards, total ban on unacceptable activities, the protection of goods of interest for society as a whole), it is not thought to be sufficient for promoting the sustainability of economic growth. This latter objective requires a correction of the market system, which is incapable of adequately accounting for environmental resources in an economic sense.

8. CONCLUSIONS

At the international level there has thus been a considerable growth in tourist activity, which has produced undeniable economic advantages, especially in the improvement in the balance of payments, employment, territorial economic equilibrium, as well as leading to an increase in income. The growth of tourist activities is nevertheless the cause of a large number of negative external effects, especially environmental ones, which, if not properly controlled, can compromise the growth possibilities of tourism itself.

Public sector intervention is thus needed to provide incentives for the development of tourism and to promote additional possibilities for increasing income, as well as to assure that development is compatible with local territorial conditions and lasts over time. An explicit economic policy is thus needed that aims at these objectives and that clearly specifies them, defining the measures to be taken, determining the priorities, evaluating the available resources, and checking that the measures are efficiently undertaken. This is also important since, though from a theoretical point of view sustainable growth has broad-based support, the concept remains an abstraction for most economic actors, who are not always able to translate objectives into everyday actions.

The measures that have been proposed, though quite different from one another, have the common aim of trying to solve the possible trade-offs between tourist activities and local territorial conditions (environmental, social, economic). This can be achieved by referring to three lines of action that concern the protection of the natural environment, the promotion of technological innovation, and in general all actions which are useful for creating a relational system among the actors operating in the territory.

NOTES

1. As it is well known, a system is a collection of entities, characterised by the interrelations among them.
2. Declaration of Manila on the World Tourism, quoted in *OCDE,* 1996, p.7.
3. See E. Becheri and M. Manente (1997), p. 6.
4. The assessment of the impact of tourism on the economy is even more complex, and cannot be dwelt here. It is sufficient to recall that a number of attempts have been made, from the OCDE's Tourist Economic Account to Tourist Satellite Accounts made by the Canadian Statistical Institute and by the World Tourism Organisations: see OCDE (1996).
5. "The most recent expressions of this phenomenon are the creation of actual 'islands' in an area profoundly changed by the preceding phases of tourist development. This is where the Bali syndrome manifests itself. Organised tourism, paradoxically, is forced to defend itself from the deterioration of the environment which tourism itself has previously given rise to. Tourism protects and distances itself from a dangerous offspring; it becomes a prisoner of itself.": C. Minca (1996), pp. 80-81.
6. See T. Leitersdorf (1995), p. 171.
7. On the danger of regional centralism, see A. Fossati (1999).

REFERENCES

Arrow K. J. (1962), "Economic Welfare and the Allocation of Resources for Inventions", in Nelson R. R. (ed.), *The Rate and Direction of Inventive Activity*, Princeton University Press, Princeton.

Becheri E., Manente M. (1997), "Economia internazionale e turismo", in *Settimo rapporto sul turismo italiano*, Turistica, Firenze

Bresso M. (1993), *Per un'economia ecologica*, La Nuova Italia Scientifica, Roma.

Butler R.W. (1980), "The Concept of a Tourist Area Cycle of Evolution: Implications for Management of Resources", *The Canadian Geographer*, XXIV, 1.

Calzoni G. (1988), *Principi di economia dell'ambiente e di gestione turistica del territorio*, Angeli , Milano.

CEE (1992), *Towards Sustainability*, COM(92)23 Final, vol. II, Brussels.

Coase R. (1960), "The Problem of Social Cost", *Journal of Law and Economics*, 3: 1-44.

Cocossis H., Nijkamp P. (eds.) (1995), *Sustainable Tourism Development*, Aldershot, Avebury.

Gerelli E. (1995), *Società Post-Industriale e Ambiente*, Angeli, Milano.

Fossati A. (1999) "Towards fiscal federalism in Italy", in Fossati A. and Panella G. (eds.) *Fiscal Federalism in the European Union*, Routledge, London,

Hicks J.A. (1946), *Value and Capital*, Oxford University Press, Oxford.

Indovina Fabris F. (1997), "La normativa nazionale e regionale", in *Settimo rapporto sul turismo italiano*, Turistica, Firenze, 1997.

Krutilla J.V. (1967), "Conservation Reconsidered", *American Economic Review*, 47: 777-786.

Leitersdorf T. (1995), "Tourism Development in the Negev", in Bar-On R.R. and M. Even-Zahav (eds.) *Investments and Financing in the Tourism Industry*, Israel Ministry of Tourism, Jerusalem.

Magnani I. (1974), "Parchi Naturali e Foreste", in Bognetti, G., Gerelli, E. (ed.), *Beni Pubblici. Problemi Teorici e di Gestione*, Angeli, Milano.

McConnell K.E. (1966), *Income and the Demand for Environmental Quality*, Seventh Annual Conference of the European Association of Environmental and Resource Economists, Lisboa.

Minca C. (1996), *Spazi effimeri*, Cedam, Padova.

Nijkamp P., Verdonkschot S. (1995), "Sustainable Tourism Development: A Case Study of Lesbos", in Cocossis H., Nijkamp P. (eds.): 127-140.

Nordahaus W.D. (1967), *Invention, Growth and Welfare*, Cambridge University Press, Cambridge

OCDE (1996), *Statistiques du Tourisme de l'OCDE. Conception et application pour les pouvoirs publics,* OCDE, Paris

Pigliaru F. (1996), "Economia e Turismo Sostenibile: alcune note", in Moro B. (ed.), *Capitale Naturale e Ambiente*, Angeli, Milano.

Pigou A. (1920), *The Economics of Welfare*, MacMillan, London.

Pearce D., Markandya A. and Barbier E. (1989), *Blueprint for a Green Economy*, Earthscan Publications, London.

Pezzey J. (1989), *Economic Analysis of Sustainable Growth and Sustainable Development*, The World Bank, WP 15, Washington D.C.

Rawls J. (1971), *A Theory of Social Justice*, Harvard University Press, Cambridge.

Sen A. (1985), *Scelta, Benessere, Equità*, Il Mulino, Bologna.

Smulders S. (1994), *Growth, Market Structure and the Environment*, Illvarenbeek.

Solow, R.M. (1986), "On the Intergenerational Allocation of Natural Resources", *Scandinavian Journal of Economics*, 88, (1): 141-149.

Tisdell A.C. (1991), "Tourism, outdoor recreation and the natural environment", in Tisdell A.C., *Economics of Environmental Conservation*, Amsterdam.

WCED (1987), *Our Common Future*, Oxford University Press, Oxford.

2. TOURISM, MARKETING AND TELECOMMUNICATION: A ROAD TOWARDS REGIONAL DEVELOPMENT

PETER NIJKAMP

1. THE GLOBAL TOURIST VILLAGE

Tourism is part of the leisure sector which is rapidly gaining economic importance. The volume of tourist flows at a world – wide level is showing a continued growth path, mainly as a result of income increase and improvement of transport systems (see Pearce, 1981). People are travelling more frequently and over longer distances for leisure purposes. Our world tends to become a global tourist village. Remote destinations are in easy reach and the modern telecommunications sector offers direct information access to such destinations.

The distribution of tourist flows over different destinations varies and is dependent on various factors (see Pearce, 1981):

- attractions   (e.g. monuments, natural parks, beaches)
- accessibility   (e.g. air connections, road infrastructure)
- accommodation (e.g. hotels, camping places)
- infrastructure   (e.g. sewage systems, telecommunications)
- suprastructure   (e.g. restaurants, banks, hospitals)

Clearly, these factors are not given, but can be influenced by local, regional or national authorities in various ways (see Burns and Holden, 1995). Examples of such facilitating and flanking policies are: the creation of a flexible legal framework (e.g. visa regulations), the provision of transport facilities, the supply of support services (e.g. clean water), the establishment of land use zoning for tourist purposes, the provision of financial and fiscal incentives, and the design of marketing tools (training of personnel, tourist information's or coordination of marketing efforts).

An active – and even pro-active – tourist policy is necessary, as the demand for tourist products in various locations shows a clear life cycle: a new tourist destination is explored and discovered, it attracts adventurous tourism, followed by mass tourism, while next it shows signs of saturation and decline (see also Prosser, 1994).

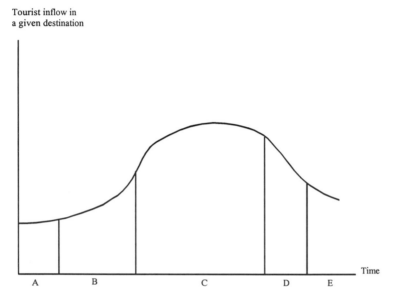

Legend: A: discovery; B: growing popularity; C: fashion and saturation; D: fading fashion; E: decline.

Figure 2.1. The tourist life cycle

Prosser argues that the dynamics in the choice of tourist destinations is caused by three motives: conspicuous consumption of an elite class, successive class interventions of lower income classes imitating the elite's behaviour, and the expansion of the tourism frontier (or "pleasure periphery"). Consequently, there are waves of tourism all over the world. This phenomena can easily be depicted in a tourist life cycle figure which maps out the various stages of tourist demand at a certain holiday destination (see Figure 2.1). This life cycle incorporates clearly distinct classes of tourists who have their own specific behaviour in the various stages of the tourist life cycle, viz. explorers, off-beat adventurers, elites, early mass and mass package tourists.

Against the previous background observations, it is plausible that, with a growing welfare and a world-wide improvement of transport systems, the demand for tourism will continue to grow and also to become increasingly differentiated. At the supply side, there will be more market specialisation and segmentation (see also Amelung, 1997).

2. THE MAGIC OF TOURISM

In the eyes of many decision-makers and politicians tourism has a magic potential. It generates income and is based on indigenous resources of the tourist areas at hand. Tourism has indeed been a rapidly growing sector and a wide-sweeping socio-economic phenomenon with broad economic, social, cultural and environmental consequences. It is likely that tourism will continue to dominate the international scene for many years to come. Nevertheless, there seems to be significant structural changes in tourist demand which are likely to influence the traditional model of "mass tourism", as the sector has been experiencing dramatic changes over the last few years in response to broader patterns of globalization of economic activities. There is growing evidence of an emerging new tourist 'profile' in connection with the change in view and behavioural patterns of all "actors" involved in planning and management of the tourist industry as well as new trends taking place not only in the area of demand but also in the interrelated area of supply (see Coccossis and Nijkamp, 1995).

In a short period of time, international tourist demand in Europe increased from 113 million arrivals in 1970 to 196 million in 1980 and to 275.5 million in 1990. It is forecasted that the growth of tourism will continue to rise to about 340 million tourist arrivals in Europe in the year 2000. This rapid increase in demand has created – and will create – several positive and negative impacts on the economy, society and environment of tourist countries and regions.

Over the past forty years tourism has become a major activity in our society and an increasingly important sector in terms of economic development. It forms an increasing share in discretionary income and often provides new opportunities for upgrading local environment. Tourism is increasingly regarded as one of the development vehicles of a region, while it is an important growth sector in a country's economy. However, much empirical evidence has also shown the negative effects of tourism, in particular on the environment. A new concept which has begun to dominate the tourism debate in recent years is that of "sustainable development" (see Giaoutzi and Nijkamp, 1994).

The idea of sustainable tourism development is now a popular concept and refers to allowing tourism growth while at the same time preventing degradation of the environment, as this may have important consequences for future quality of life. In this context, Buhalis and Fletcher (1992) quote Goodall who has suggested that sustainable tourism requires that "the demand of increasing numbers of tourist is satisfied in a manner which continues to attract them whilst meeting the needs of the host population with improved standards of living, yet safeguarding the destination environment and cultural heritage".

Tourism is thus not only a rapidly rising economic activity, on all continents, in countries and regions, but it is also increasingly recognized that this new growth sector has many adverse effects on environmental quality conditions. In the context of the world-wide debate on sustainable development there is also an increasing need for a thorough reflection on sustainable tourism, where the socio-economic interests of the tourist sector are brought into harmony with environmental constraints, now and in the future. Tourism is intricately involved with environmental quality, as it affects directly the natural and human resources and at the same time is conditioned by the quality of the environment. Such a relationship has important implications from the point of view of policies, management and planning.

Tourism is thus a double-edged sword. It may have positive economic impacts on the balance of payments, on employment, on gross income and production. Also, tourism development may be seen as a main instrument for regional development, as it stimulates new economic activities (e.g. construction activities, retail shopping) in a certain area. Nevertheless, because of its complexity and connection with other economic activities, the direct impact of tourism development on a national or regional economy is difficult to assess. Clearly, a careful assessment of the environmental impacts of tourism is very important, because tourists tend to be attracted to the more fragile environments, for example, small islands, centres of high historical and cultural value, and coastal zones. Tourist development thus poses special problems for environmental resources which are 'exploited' by tourism. The use of such environmental resources for tourism has two consequences. The quantity of available resources diminishes and this, in turn, limits a further increase of tourism. Besides, the quality of resources deteriorates, which has a negative influence on the tourist product.

Tourism and the environment are thus interdependent. The environment is one of the most important factors in the tourist product, as the quality of this product depends on the quality of the environment, which is the basis for attracting visitors and hence has to be conserved. Tourist development depends then on a proper handling of this close relationship between tourism and the environment. Therefore, it is necessary to examine both the regional economic and environmental impacts of tourism with a view on the implementation of a balanced tourist policy. This will be the subject matter of Sections 3 and 4, respectively.

3. ECONOMIC IMPACTS OF TOURISM

It goes without saying that tourism induces changes in many areas, not only socio-economic, but also environmental. In this section we will address in particular the effect on the national and regional economy. The assessment of such effects is however, fraught with many difficulties, as the tourist sector comprises a complex set of interlinked activities, such as travel, accommodation, catering, shopping etc. (see for details Briassoulis, 1995). Consequently, in a strict sense tourism cannot be considered as a specific sector or industry; it is essentially a complex ramification of economic activities which in combination determine the quantity or quality of the tourist product of an area. Thus, given the multi-activity and multi-sectoral nature of tourism, the sectors constituting tourism contribute differently to the production and consumption of the tourist product. Firms in the tourist sector are numerous, offering various types of product and each one contributing to the quality of the tourist product. This product is thus a "packaged" selection of elements which are decisive for the perceived attractiveness of a tourist place. This also means that the determinations of a single unambiguous price for the tourist product is almost impossible.

Consumers face a multi-product and multi-priced set of amenities, on which they do not have perfect information and free choice, so that they tend to take sub-optimal decisions. And finally, the tourist product is a mix of private and public goods which complicates the application of purely public or purely private sector policies for the control of its quality and impacts. An additional complication is that the tourist sector has the typical features of a seasonal activity which may lead to discontinuous economic impacts. In summary, the tourist demand is dynamic, fluctuating over space and time as a result of frequent changes in tourist preferences and marketing policies.

The socio-economic effects of tourism are manifold and can be classified as follows (see Pearce, 1991):

- balance of payments: tourism is essentially an export good which brings in foreign currency, although foreign tourist operators, promotion campaigns abroad etc. may reduce the net benefits for the balance of payments;
- regional development: tourism also addresses peripheral areas and hence spreads economic activity more evenly over the country;
- diversification of the economy: given the multi-faceted nature of the tourist sector, it may help to build up a robust economic development;
- income levels: the income effects of tourism are twofold: building / construction and operations. This may also explain variation in income multiplies (see later);
- state revenue: the state earns revenues due to tax collection, although it has to be recognized that also significant outlays for the infrastructure may be needed;
- employment opportunities: tourism is rather labour – intensive and requires also much unskilled and semi-skilled labour, which offers great opportunities for less favoured regions.

The extent to which these effects will manifest themselves varies a lot and is dependent on the stage of the tourist life cycle, local tourist policy strategies and the use of sophisticated communication technology in promotion campaigns. In all cases, the quality of the tourist product offered is decisive for the economic impact on the local a regional economy. In this context, a keen marketing strategy is of utmost importance, as such a strategy has to ensure the best possible match between the tourist's aspiration level and the opportunity set of the tourist's resources and attributes.

4. ENVIRONMENTAL IMPACTS OF TOURISM

The benefits of the tourist sector have to be compared with the social cost imposed by this sector. The environmental consequences of tourism are often manifold; they do not only manifest themselves in terms of noise or pollution, but also in terms of destruction of landscape, a decay in the quality of cultural heritage, or even a destruction of local communities. Several of there impacts are irreversible, while in many cases the social costs are not charged to the tourist. Thus, tourism is a typical example of an industrial sector which depends on natural and cultural resources and which threatens to erode the basis of its activities (Bramwell et al.,1995). According to

Buckley (1994), there are four links between tourism and the environment:

- components of the natural environment as the basis for a marketable tourism attraction or product;
- management of tourism operations so at to minimize or reduce their environmental impacts;
- economic or material contribution of tourism to conservation;
- attitude of tourist towards the environment and environmental education of clients by tourist operators.

Given the multidimensional dynamic complexity of the tourist sector, it is not easy to identify a sustainable development path for the tourist sector. There are conflicting objectives involved, and the definition of an unambiguous sustainable state for tourism is a thorny question. Müller (1994) has made an attempt to specify a sustainable tourism development by using a so-called "magic pentagon" (see Figure 2.2). This magic pentagon takes for granted that sustainable tourism reflects a state of affairs where economic health, the well-being of the local population, the satisfaction of the visitors/tourists, the protection of the natural resources and the health of the local culture are in balance. Any imbalance in this prism means a distortion and will negatively impact the benefits of all acts involved.

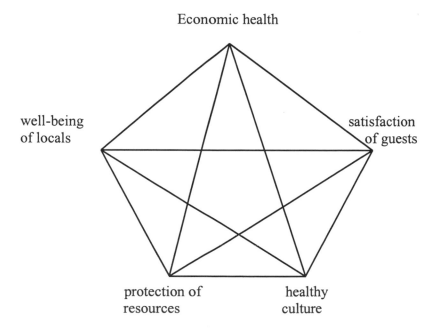

Figure 2.2. The 'magic pentagon' of sustainable tourism

It is clear that this figure is merely indicative and does not offer an operational definition. Eber (cited in Bramwell et al., 1996, p. 39) defines sustainable tourism as "tourism which operates within natural capacities for the regeneration and future productivity of natural resources; recognizes the contribution that people and communities, customs and lifestyles, make to the tourism experience; accepts that these people must have an equitable share in this economic benefits of tourism; and is guided by the wishes of local people and communities in the host areas". Naturally, the achievement of a state of sustainable tourist development is fraught many difficulties, causes by the high pressure exerted by the tourism demand, the specific, egocentric nature of tourism, and the look of ecological awareness in the tourist sector. Sustainable tourism policy challenges will be discussed in Section 5.

5. TOWARDS SUSTAINABLE TOURISM POLICY

One of the most important prerequisites for attracting tourists of an area is the beauty of the natural environment. But if such an attraction force becomes successful, it has to be recognized that too large a number of tourists will damage or even destroy the natural environment, thus eroding the very basis of tourist activity. Several examples clearly illustrate this observation, like the case of Venice: a further increase of tourists in Venice will negatively influence the tourist attractiveness. This provokes of course the question of *carrying capacity* (and related policy strategies) of tourist areas (see Ceballos-Lascurain, 1996; Weaver, 1998).

It is possible to avoid a situation in which the natural environment is damaged as a result of severe negative external effects from tourism? Several policy instruments can be envisaged to support the implementation of sustainable tourism. We mention the following possibilities:

1. Information and education from the side of the government is mainly meant to make citizens, tourists and companies aware of the environmental problems and their role in it; it serves to show how they can contribute to the solution of environmental problems. It may also increase public support for government policy. An example is information on the amount of waste tourists produce (for the tourists) and the costs they impose on society (for the local inhabitants).
2. Subsidies belong to the group of economic instruments; it is a market-oriented instrument. This means, for instance, that it is not prohibited to

pollute, but policy activities will be (relatively) more expensive. The government can in this way stimulate environment-friendly behaviour (for instance, by imposing lower prices on tourist facilities in ecologically less vulnerable areas). Taxation is a similar instrument and tries to discourage environment unfriendly behaviour. The government can impose a tourist-tax ("tourist pricing") to discourage tourism in a specific area.

3. Legal instruments can have many forms. One of them is liability; this means that companies and citizens are liable (responsible) for damage they (or their products) cause to the environment. It is even possible to use this instrument in such a way, that companies have to prove that their products are not harmful to the environment. Hotels, for example, might have to prove that the tourists they receive (or their tourist facilities) do not cause any damage to the environment.

4. The government may, for example, supply public infrastructure like waste treatment facilities, but they can also supply an infrastructure to make it easier for the private sector or tourists to act more environment benign (e.g.,supply public transport to tourist attractions). However, it is of course not certain that they will actually use it. This type of infrastructure can be supplied by the private or the public sector but also by public-private partnerships (PPPs) for efficiency reasons.

5. The government can also make agreements about e.g. the reduction of pollution. This instrument is more flexible than strict regulations or quota. It is however, important to point out that the government can always resort to laws, if companies are not willing to act as agreed upon (to prevent the "prisoners dilemma"). For example, the maximum number of hotel rooms or the maximum height of hotels can be agreed on. This instrument is especially useful in "prisoners-dilemma" situations, which often occur in the tourism development.

6. Permissions or quotas for the amount that a company is permitted to pollute are instruments which give the government the opportunity to determine the exact amount of future pollution. With subsidies or taxation, this is not always possible. The problem of permissions is that it can lead to market imperfections and illegal actions. A way to decrease this problem is by making the permission tradable: the permission can be sold by means of an "auction". The maximum number of tourists can also be fixed in this way ("tourist quota").

7. There may also be a free market strategy. Free market means here that there is no direct intervention in the market system by the government. The question is of course whether and how external costs of tourism can

be internalized or computed.

8. Finally, there may be a niche strategy. Niches are specific fractions of the total market. A group of suppliers of tourist facilities can, for example, decide to concentrate on a certain segment of the tourist market, like nature tourism or health tourism. Such niche market policies will be discussed hereafter.

6. MARKET NICHES OF TOURISM

Niche markets are a well-known phenomenon in the tourist sector. They tend to exploit the competitive advantages of specific market segments. For many tourist markets such as the Greek one, the following niches can be distinguished:

Exclusive tourism

Exclusive or top-class tourism is aimed at the arrival of high income tourists. A limited, rather exclusive market, which will not require a further extension of the present built up area, is needed. Exclusive tourism should first be implemented in the main tourist places, which have most "B" level hotels. These existing hotels may first be upgraded to a higher category, "A" or "first class", through the improvement and addition of more (luxury) facilities, better services provision and higher standards of cleanliness. These improved standards of quality should also be applied to restaurants and other supporting facilities.

Agri-tourism

Agri-tourism is a kind of tourism which favours the economic activities in the agricultural sector at the same time. An important aim is to stimulate these activities in relation to the agricultural potential so that the economy of the region will not become solely dependent on tourist activities. Agri-tourism contains, for instance, the construction of tourist accommodations and facilities at farmers' places. Besides, tourists may watch the processing of farm products. Olive oil, ouzo production and leather industry are popular traditional activities to which agri-tourism may also be applied. Agri-tourism can be implemented in many rural areas.

Health tourism

The area may also develop facilities for curative tourism. For example, thermal waters (which are present in various islands) are recommended for people with rheumatic problems, bronchitis, back aches, skin diseases etc. Bath facilities, accommodations around the spa's and access roads may then be improved or established. Medical tourism seems to become an increasingly important part of the tourism market.

Adventure/sports tourism

Many regions lend themselves for specialisation in sport and leisure tourism. Several Greek Islands are ideal for sports thanks to their nature and culture. For example, the National Tourist Organisation has established different trekking routes in various areas. An information guide with the different trails is also published by this organisation. Such pathways need to be better cleaned, mapped, marked and developed. Another possibility is the organisation of wildlife/discovery tours, so that organised tourist groups can make panoramic trips and see the countryside by walking, climbing and hiking. Other examples are bird watching, camping, horse-riding, cycling, golf or tennis. Proper facilities may be developed and constructed.

Sea tourism

Greece has a strong comparative advantage in sea tourism, because it has an abundance of surrounding waters. The sea lends itself to wind surfing, water skiing, snorkelling and sailing. Establishment of modern water sports facilities may be established at some tourist resorts. Avoidance of already saturated places would be better. For yachting, the construction of modern marina's with an environmental control would be desirable.

Cultural tourism

Greece has a great variety of typical cultural, historic and natural attractions. By upgrading the level of and access to these attractions, the country will gain cultural prestige and may offer more interesting places to visit. The level of service and quality of museums might be improved. Organisation of art exhibitions or other cultural manifestations may also be an interesting possibility. The typical traditional villages also require protection. The materials used and the design of buildings in such places may be based on local traditions. The exercise of cultural handicrafts might be encouraged.

Winter tourism

Winter tourism may be encouraged to realise a year round tourist product. This is also related to targeting winter migration to the island by offering facilities to elderly people. An easily attainable island in the winter months and the presence of qualified tourist services are a prerequisite. An advantage of winter tourism is the creation of jobs in winter time, so that seasonal unemployment is reduced.

Education tourism

Tourism may also be developed on the basis of meetings, conferences congresses or symposia. The construction of congress centres would then offer a new opportunity. Organisation of language or cultural courses may also be considered.

These various tourism options seem to be the most feasible ones and do not exclude each other. The development of a mix of different options is thus possible and desirable.

Environmental instruments in the tourist sector should be designed to facilitate the integration of environmental policy with other policies, such as regional development policies. Removal and correction of administrative and governmental intervention failures are therefore of importance for a proper integration of environmental policy with sectoral policies. This may end up in a better synergy and coordination of tourist activities with other socio-economic activities which will be outlined in the next section.

7. TOURISM, TECHNOLOGY AND THE REGION

It is a well-known fact that the revenues in the transportation sector – an activity meant to physically transport people and goods – are rather modest. In fact, the transportation sector – interpreted in the above mentioned limited sense – is just like the agricultural sector a infra-marginal economic activity. However, in a modern economy the strategic importance of the transportation sector and the supra marginal profitability of parts of this sector are dependent on the logistic management of the sector. In other words, the physical movement is not the main source of revenues, but the non-physical organisation and coordination of the sector based on modern telecommunication, telematics and logistic services. The latter branch of economic activity does not constitute a low-skilled segment of the labour market, but is determined by highly educated, specially trained and internationally-oriented employees. Thus, distribution has become more important than material production and transportation.

The same observation applies to other economic sectors, like agriculture, banking, computer manufacturing and repair services. The "deskilling" hypothesis which used to be fairly popular in the 1980s has turned out to be wrong: in all sectors the highest value added is created in the upper segments of the labour market.

This observation applies also to the tourist sector. This sector started as a simple, relatively low-skilled segment of the market by offering accommodation and related services to travellers. This traditional picture has drastically changed in the past years. First, the economic prosperity has created a relatively large share of leisure time and discretionary income, so that more people could enjoy the benefits and pleasure of international tourism. Second, the world-wide mobility movement has drawn the attention of potential customers to distant and unknown destinations. And finally, modern telecommunication means bring attractive tourist destinations directly in the living rooms of potential travellers. This means that welfare rise and modern information and communication technology may be held responsible for the global drive towards mass tourism.

At the same time a drastic restructuring of the tourist industry itself has taken place. First, a concentration in the sector has occurred, witness the emergence of large scale international hotel chains. Secondly, as a result of those economies of scale a further rationalisation has taken place, where electronic booking and advanced pay systems have taken over the role of the conventional handicraft nature of the hotel business. But more importantly, the organisation of tourism has come in the hands of a few large-scale tour operators who govern a significant part of the international market. These operators form a critical intermediate segment between demand and supply they do not only organize packages of trips for the traveller, but they dominate increasingly the hotel accommodation market as well as the tourist transport market. By a keen combination of various opportunities and by using the modern information and communication technology as a spearhead, they are in a position to control large parts of the travel agency market and the transport market for tourists. As a consequence, both tourists carriers and hotel owners tend to become increasingly dependent on vested interest of a highly qualified and technologically well developed group of tour operators. The question which may be raised now is: how can regions exploit the opportunities of the modern information and communication sector, without falling in the hands of the monopoly power of international tour operators?

It seems thus that a package of measures can be envisaged to nurture indigenous strength and to seek for cooperation at the regional level. Elements of such a package are:

- increase the quality of supply of tourist facilities by addressing in particular the environmental quality of the area;
- coordinate the information on the supply of tourist facilities (hotels, but also culture and nature) at the regional level, so that the region can be conceived of as the supplier of a strong package of attractive tourist facilities;
- invest in sophisticated regional information and communications technologies (e.g., on electronic booking systems, internet information on the area via a website page);
- organize the regional forces so that a uniform tourist image of the area is shaped for the international traveller which may favour an inflow of tourism without being dependent on international tourist operators.

Thus there is a main role for marketing the tourist product by creating new customers through a balanced combination of product, price, distribution and communication services, as such services are often more effective and efficient than other forms of marketing. It is recognized in the tourist sector that markets gain in competitive advantage with improved communication, due to better information access and distribution as well as a more proper response to market signals.

It is clear that the tourist sector may be a source of important revenues, but a significant part of these revenues is often lost for tourist areas, because either the owners of tourist facilities live outside the area or the tourists pay their trip and accommodation via tour operators abroad. It is also noteworthy that the tourist income multipliers show a formidable range of variation among different tourist areas. It seems plausible that the organisation of the tourism market is to a significant degree responsible for these differences. This will be further illustrated and analysed in the next section.

8. ASSESSMENT OF REGIONAL TOURIST INCOME MULTIPLIERS

International tourism may broaden and deepen the supply side of an economy as a result of the additional and probably more diversified demand generated by it. Impacts of the tourist sector on the various domestic economic sectors can be subdivided into (stimulating) effects on production, gross income and employment. In as far as more sectors benefit from activities in the tourist sector, the notion of multiplier effects initiated by incoming tourist expenditures is relevant. A multiplier value may be interpreted as a stimulus-response ratio of effects vis-à-vis the initial (monetary) injection. The quantification of these effects by means of so-called tourist multipliers is a modification of the standard Keynesian multipliers ("snowball effects"), developed in a general context, for the tourist sector. The magnitude of these multiplier effects is determined by the way in which initial tourist receipts filter throughout the economy stimulating linked sectors on their way. Tourism demand is met by the output of tourist sectors, which again require deliveries from linked sectors and so forth.

It is without any doubt true that tourism is one of the world's fastest growing industries. Regional impact assessment of tourism is usually based on multiplier analysis which aims to depict all (direct, indirect and induced) consequences of additional tourist expenditures (see Nijkamp and Baaijens, 1998). The chain of effects after an initial impulse is often described by means of an input-output model, incorporating all transactions between relevant economic sectors. The environment (and, in general, the resource sector) can also be included as one of the inputs or outputs related to the production of goods and services. In this way, the system-wide impacts of tourist expenditures in a regional economy can be traced (see also Armstrong and Taylor, 1993). A more simplified approach to the estimation of regional tourist multipliers has been developed by Archer (1976) who made a distinction between tourism-aligned and non-tourism-aligned sectors. But even in that case the calculation of multipliers is a tedious and expensive task. We will emphasize here the need to undertake cross-national comparative studies on the values of tourist income multipliers on the basis of principles of meta-analysis.

The start of a meta-analytical exercise is to collect documented studies on income multiplier effects of tourist areas, notably regional economies. In a recent study (see Nijkamp and Baaijens, 1998), we were able to identify a sample of 11 relevant and officially published case studies on tourist regions from different sources and covering different years (see Table 2.1). The information provided by the study reports served as the principal data base in the meta-analysis. Clearly, not all relevant data were directly available for a comparative quantitative analysis as several data was more or less hidden in the study reports. Therefore, a systematic inventory of all relevant information had to be made. This investigation was made with the help of a systematic list of relevant topics, by making also use of a general framework of economic meta-analysis (see for details van den Bergh et al., 1997). A closer investigation of the study reports led to a systematic overview of the relevant information contained in the study reports. This overview however, appeared to be far from complete, as each individual study contained differences in approach and viewpoint, so that the similarity in the information provided left much to be desired. As a consequence, the meta-analysis was restricted to information that was known for most geographical areas considered. The final selection of the cross-sectional case-study features is contained in Table 2.1. Ideally, one would wish to have more specific economic variables, such as the value-added ratio or the import ratio, or more specific tourist information such as tourist days or purchasing power parity data. Unfortunately, these data were not available to a sufficient extent in the study reports to warrant a solid meta-analysis. Clearly, missing information limits the use of cross-sectional comparisons.

The range of variation in the multiplier values appears to be rather significant. Consequently, it does not make sense to make an average estimate of such a multiplier of any new case study. Rather, it is necessary to link the value of these multipliers to differences in background variables using meta-analytical techniques. This meta-analytic experiment led to quite some interesting results.

Clearly, the application of statistical meta-analytical techniques for the assessment of the tourist income multiplier incorporates significant differences with the application of conventional meta-analytical techniques to studies which deal with a controlled experimentation model. In our case of 11 tourist studies, we do not have a controlled experiment on sampling among households, tourists and other economic sectors in the regional economy (the studies are given). Furthermore, the tourist analysis is carried out on a meso level of aggregation, and finally, in these studies the (questionable) assumption is made that the measured (or estimated) multiplier (via the input-output or the Archer model) is also the real one and not subject to stochasticity.

We will carry out regression experiments in three steps:

- a base model where tourist multipliers are related to geographic characteristics such as surface and population. The area and the population size of the region serve as indicators for the diversity of economic activity in the region.
- a meta model where the typical meta variables are included in the explanatory analysis, e.g., the source of the study, the year of collection

of data, the analysis method used. Meta variables refer to methodological choices made in the original studies; their values are largely independent from the object of the study in the meta-analysis.

- a tourist model where the size of the multiplier will be linked to incoming tourist flow

Base model

In the base model the tourist multiplier is, in general, positively correlated with the (natural logarithm of the) size of the population. In the base model the tourist multiplier is, in general, positively correlated with the (natural logarithm of the) size of the population.

Next, also the size of the area has been investigated. The regression results appear to support the hypothesis that regions with a larger surface have a higher tourist multiplier.

Finally, the impact of the degree of political autonomy has also been examined. In contrast to our hypothesis, we find a significant negative relationship between the income multiplier and political autonomy.

Meta model

The meta model has been experimented in combination with population size. We find that the impact of population size in the meta model is again significantly positive that estimates of multipliers, published in scientific journals, appear to be lower than those published elsewhere; there is apparently, on average, a tendency towards some overestimation in less official publication channels.

Next, it is also noteworthy that the type of model used has an impact on the results: the Archer model tends to yield, on average, lower values of the tourist income multiplier than the full input-output model.

Tourist model

In the final stage we investigate the impact of tourism-specific variables, in addition to the previous significant variables of population size. There is some indication that higher tourist arrivals lead to a higher multiplier value. The impact of the tourist index (ratio of incoming tourists to population size) appears to be positive.

It should be noted that the results are merely indicative, but nevertheless interesting, as they generate some plausible ideas on the impacts of base, moderator and tourism-specific variables, which may be transferred to other situations.

Table 2.1. Concise survey table for meta-analysis for tourist areas

	DOC	GEO	YEA	REM	TIM	ATR	POP (x 1000)	SUR (x 1000)	TOA	LCS	POA
Bahamas	D_r	G_g	1976	R_a	0.7815	A_s	189.9	11,401	1388.0	75.8	y
Bermuda	D_r	G_g	1975	R_a	1.0996	A_s	56.6	107	511.4	86.6	n
Singapore	D_j	G_i	1983	R_{io}	0.9393	A_c	2501.0	625	2856.6	?	y
Turkey	D_j	G_c	1981	R_{io}	1.9809	A_m	45,529.0	779,425	1,460.0	?	y
Niue	D_j	G_i	1987	R_a	0.35	A_n	2.0	258	1.8	78.0	n
Cook Islands	D_j	G_g	1984	R_a	0.43	A_s	18.0	236	25.6	40.0	n
Kiribati	D_j	G_g	1987	R_a	0.37	A_n	66.0	270	2.0	?	y
Tonga	D_j	G_g	1987	R_a	0.42	A_s	95.0	699	16.1	24.2	y
Vanuatu	D_j	G_g	1987	R_a	0.56	A_m	140.0	12,200	17.5	52.0	y
Alonnisos	D_b	G_i	1989	R_{io}	0.489	A_s	1.55	83	20.0	35.9	n
Okanagan	D_j	G_r	1977	R_{io}	0.713	A_n	?	21,813	1,400.0	?	n

Legend:

1. Type of documentation (DOC):
   D_r: research paper
   D_j: journal
   D_b: book

2. Geographic feature of the area (GEO)
   G_s: single island
   G_g: island group
   G_r: region
   G_c: country

3. Year of collection of data (YEA)
4. Research method (REM):
   R_{io}: input-output model
   R_a: Archer method
5. Estimated value of the average tourist income multiplier (TIM)

6. Type of tourist attractiveness (ATR):
   A_s: sun
   A_c: culture
   A_n: nature
   A_m: mixed
7. Quantitative and qualitative features of the tourist area:
   Population size (POP)
   Surface (in km^2) (SUR)
   Tourist arrivals (TOA)
   Share of arrivals from most important country of origin in total number of arrivals (LCS)
   Political autonomy (POA)
8. Population size (POP)
9. Surface (in km^2) (SUR)
10. Share of arrivals from most important country of origin in total number of arrivals (LCS)
11. Political autonomy (POA)

REFERENCES

Amelung S.B. (1997), *Tourism, Environment and Policy in Costa Rica*, Master's Thesis, Dept. of Economics, Free University Amsterdam.

Armstrong H. and J. Taylor (1993), *Regional Economics and Policy*, Philip Allen, Oxford.

Archer B.H. (1976), The Anatomy of a Multiplier, *Regional Studies*, 10, (1): 71-76.

Bergh J.C.J.M. van den, K. Button, P. Nijkamp and G. Pepping (1997), *Meta–Analysis in Environmental Economics*, Kluwer, Dordrecht.

Briassoulis H. (1995), "The Environmental Internalities of Tourism", in H. Coccossis and P. Nijkamp (eds.), *Sustainable Tourism Development*, Avebury, Aldershot: 25-40.

Briassoulis H. and. J. van der Straaten (1997), *Tourism and the Environment*, Kluwer, Dordrecht.

Bramwell B., G. Jackson and J. van der Straaten (1996), *Sustainable Tourism Management: Principles and Practice*, Tilburg University Press, Tilburg.

Buhalis D. and J. Fletcher (1995), "Environmental Impacts and Tourist Destinations; An Economic Analysis", in H. Coccossis and P. Nijkamp (eds.), *Sustainable Tourism Development*, Avebury, Aldershot: 3-24.

Buckley, R. (1994), "A Framework for Ecotourism", *Annals of Tourism Research*, 21, (3): 234-252.

Burns P. and A. Holden (1995), *Tourism: a New Perspective*, Prentice–Hall, London.

Ceballos-Lascurain H. (1996), *Tourism, Ecotourism, and Protected Areas*, IUCN, Gland, Switzerland.

Coccossis H. and. P. Nijkamp (eds.) (1995), *Sustainable Tourism Development,* Avebury, Aldershot.

Giaoutzi M. and P. Nijkamp (1994), *Decision Support Models for Regional Sustainable Development*, Avebury, Aldershot.

Müller H. (1994), "The Thorny Path to Sustainable Tourism Development", *Journal of Sustainable Tourism*, 2, (3): 106-123

Nijkamp P. and S. Baaijens (1998), "Meta-analytic Methods for Comparative and Exploratory Policy Research", *Journal of Policy Modeling* .

Pearce D. (1981), *Tourist Development*, Longman, New York.

Prosser R. (1994), "Societal Change and Growth in Alternative Tourism", in E. Carter and G. Lowman (eds.), *Ecotourism, a Sustainable Option?*, John Wiley, Chichester: 89-107.

Weaver D.B. (1998), *Ecotourism in the Less Developed World*, CAB International, Wallingford, UK.

3. WHY ARE TOURISM COUNTRIES SMALL AND FAST-GROWING?

ALESSANDRO LANZA AND FRANCESCO PIGLIARU

1. INTRODUCTION

Having grown faster than world GDP since the 1950s, international tourism is today one of the most important tradable sectors, with expenditure on tourist goods and services representing some 8% of total world export receipts and 5% of world GDP.

In spite of this, the importance of this sector for a country's overall growth performance has often been neglected in economic literature. For instance, in the recent impressive survey on the economic analysis of tourism [Sinclair (1998)], only very few papers out of the hundreds reviewed deal explicitly with the long-run growth consequences of specialisation in tourism. Consequently, not much work can be found on theoretical models aimed at explaining some interesting empirical findings characterising the relationship between economic growth and tourism specialisation in cross-country data.

Consider, for instance, the following experiment in which two separate lists of countries from the World Bank data set[1] are used. The first list includes the top fifteen fastest growing countries in per capita income, from 1985 to 1995.[2] The second includes the fifteen countries with the highest degree of specialisation in tourism, defined as share of international tourism receipt with respect to the value added[3]. By simply comparing these two lists we find that seven out of the fifteen "tourism" countries appear in the list of the fifteen fastest growing ones – namely, St. Kitts and Nevis (per capita income average growth rate 1985-95: 5.9%), Singapore (5.4%), Antigua and Barbuda (5.3%), Maldives (5.1%), Mauritius (5.1%), Seychelles (4.5%) and Cyprus (4.5%). While this evidence is far from being conclusive about the effects of tourism specialisation on growth, it is enough at least to show that tourism can make a country grow fast.

The above exercise unveils a second peculiarity. All fifteen countries with a high degree of specialisation in tourism share a rather evident feature: they are *small* countries[4]. This suggestive evidence points to the likely existence of two empirical regularities which might characterise the tourism sector when viewed from a macroeconomic perspective: *i*) tourism specialisation can make a country grow fast; *ii*) countries specialised in tourism are generally small ones. While more evidence is clearly required to validate the status of "empirical regularity" for these two findings, the evidence discussed so far is enough, in our opinion, to prompt a simple research agenda aimed at generating a joint explanation of points *i*) and *ii*).

The rest of the paper is divided into two parts. Section 2 considers the relation between tourism specialisation and economic growth. The main objective of this section is to illustrate the conditions under which tourism specialisation is not detrimental to economic growth. To this aim, we use a simple endogenous growth theory framework based on Lucas (1988). Once this step is accomplished, we address the second question concerning the dimension issue. In Section 3 we discuss two alternative explanations of point *ii*), and assess their consistency with respect to the hypothesis put forward to explain point *i*).

2. TOURISM SPECIALISATION AND ECONOMIC GROWTH

Point *i*) above indicates that tourism specialisation might not be detrimental to economic growth. While obviously no sector is "detrimental" to growth in an exogenous growth setting, things may be very different when the growth rate is endogenously determined. Much of the recent literature in this field points to the positive key role the more innovative sectors play in such determination. Considering countries in isolation, a larger innovative sector may spur faster growth in the long run. If trade induces different countries to specialise in sectors with different dynamic potentials, and technological spillovers across sectors and countries are not strong enough, then uneven growth is normally obtained [Grossman and Helpman (1991); more recently, Aghion and Howitt (1998)].

While these preliminary remarks may not sound too promising for countries specialising in tourism, the endogenous growth setting is nevertheless the one we need to consider in order to address fact *i*). In particular, Lucas's (1988) two sector endogenous growth model is simple and detailed enough for our purpose of finding the conditions under which tourism specialisation is not a growth-damaging option. These conditions are discussed at length in Lanza and Pigliaru (1994) and (1995). Here we briefly summarise the main thrust of the argument, so to ease the forthcoming discussion about fact *ii*), in which we use the same formal framework.

Consider a two-sector world in which the engine of growth – the accumulation of human capital – takes the exclusive form of learning-by-doing, so that pure competition prevails. The technology to produce sectoral outputs y_i is as follows:

$$y_i = h_i L_i \qquad (i = 1,2) \tag{1}$$

where h_i is human capital, which determines labour productivity in the sector, and L_i is the labour force allocated to the sector. For the time being we assume that all existing countries have the same size of the overall labour force ($L=1$). This assumption will be dropped in the next section. In each sector the potential for learning-by-doing is defined by a constant, λ_i. In our case, manufacturing (M) is the "high technology" sector, so that $\lambda_M > \lambda_T$, where T stands for tourism[5]. In each period, with knowledge accumulation driven by learning-by-doing with external economies linking all firms within the same sector, and no intersectoral spillovers[6], the increase in h is simply proportional to the sector's output [see (1)], so that:

$$\frac{\dot{h_i}}{h_i} = \lambda_i L_i . \tag{2}$$

International trade will force all countries to specialise completely according to the comparative advantage they have when trade opens up (on this, more below, in section 3). The growth rate of a country therefore depends on such complete specialisation, according to:

$$\frac{\dot{y_i}}{y_i} = \lambda_i . \tag{3}$$

Therefore, productivity grows faster in countries specialised in M (measured in terms of this good) than in the other countries (measured in terms of T). However, with preferences assumed to be homothetic and identical everywhere, the terms of trade move in favour of the slow-growing good, tourism, at a constant rate. With CES preferences the rate of change of $p \equiv p_T/p_M$ is equal to $(\dot{y}_M/y_M - \dot{y}_T/y_T)\sigma^{-1}$, where σ is the elasticity of substitution. With complete specialisation, therefore:

$$\frac{\dot{p}}{p} = \frac{\lambda_M - \lambda_T}{\sigma}. \tag{4}$$

Comparing now the growth rates associated with the two available patterns of specialisation in terms of a common good (M, for instance), we find that tourism is the growth-maximising specialisation if

$$\frac{\lambda_M - \lambda_T}{\sigma} > (\lambda_M - \lambda_T), \tag{5}$$

that is, if $\sigma < 1$. In words, tourism is not harmful for growth if the international terms of trade move in it fast enough to more than offset the difference in sectoral productivity growth. For this to happen, the two goods must not be close substitutes.

The empirical value of the elasticity of substitution between manufactured goods and tourism is therefore an important piece of the evidence when it comes to evaluating the long-run consequences for an economy that specialises in tourism. Using an OECD countries data set, Lanza (1997) finds that in most cases σ is indeed lower than one.

So far, we have defined an explanatory hypothesis about point *i*) above. The aim is now to extend the underlying model in order to obtain an explanation of the second empirical regularity – that is, tourism countries are generally small – consistent with the explanation put forward for the first one.

3. COUNTRY SIZE, TOURISM AND GROWTH

Before turning to our own hypothesis about point *ii*), we discuss a simple and appealing hypothesis suggested by Cellini and Candela (1997). In their paper, the two authors adopt the Lucas based explanation of point *i*) as first suggested in Lanza and Pigliaru (1994), and try to extend it to address point *ii*). Our aim here is to show that, however appealing, their hypothesis is inconsistent with their more general aim – that is, to obtain a joint explanation of the two regularities within the same theoretical framework.

3.1. Explanation one: Does a country's absolute size matter?

Recently, Candela and Cellini (1997) (CC hereafter) have provided an explanation based on the idea that countries with different absolute (population) sizes face different opportunity costs associated with tourism specialisation.[7] More precisely, they note that, within Lucas's framework, "the smaller a small-economy is, the easier the pattern of the terms of trade offsets the technology gap disadvantage", so that "the opportunity cost of specialisation in tourism is smaller, the smaller is the country" [CC (1997), p. 457][8]. This point can be shown easily by assuming an exogenous constant positive growth rate of the terms of trade $p \equiv p_T/p_M$. Let us consider small countries of two different sizes, i and j, and assume that $L_j > L_i$. As CC point out, the growth maximising choice of each country facing an exogenous $\dot{p}/p$ depends on its size. For instance, if $\dot{p}/p > (\lambda_M - \lambda_T)L_j$, countries of the j type find it optimal to specialise in tourism. If so, countries of the i type would also be specialised in tourism. For them, tourism is clearly an even more convenient option than for j countries.

Implicitly, this example supports CC's major point: when countries are of heterogeneous absolute size, *if only one group of them is willing to specialise in tourism, this group must be the one formed by the countries of the smallest size.* Indeed, a value of $\dot{p}/p$ exists such that:

$$(\lambda_M - \lambda_T)L_i < \dot{p}/p < (\lambda_M - \lambda_T)L_j .$$

Here i countries are optimally specialised in tourism, and j countries in manufacturing. However, a crucial joint prediction is also revealed by this example – a rather troublesome one. Larger countries grow faster than smaller ones; therefore, countries specialised in manufacturing grow faster than tourism countries.

As point i) above suggests, the available evidence does not support these predictions. Countries specialised in tourism are small *and* they grow fast – faster than the others, on average. Countries specialised in tourism experienced a per capita income growth of about 5% per year during the period 1985-1994. During the same period the average growth rate at the World level was equal to 0.8%, and for the subset of developing countries it was equal to 3.6% [mainly due to China's high performance during the period: 8.3%].

In other words, extending Lucas's approach by introducing heterogeneity in absolute size across countries does generate an explanatory hypothesis of point *ii*). However, the same modification makes the model incapable of jointly explaining point *i*). Analytically, it is not difficult to single out the source of the problem. To address point *ii*), CC rely on a strong and generally regarded counterfactual scale effect based on the size of the labour force. As a consequence, the smaller countries' growth path is the slower one[9].

In the following, we propose an alternative explanation in which a resource-based comparative advantage is what characterises the heterogeneity across the existing countries. In our proposed explanation, we abstract again (as in section 2) from differences in absolute size in order to get rid of the scale effect discussed above.

3.2. Explanation two: Do relative resource endowments matter?

Most of the small fast growing countries specialised in tourism ground their supply of tourism services on their natural resources. This suggests a second hypothesis – alternative to the one based on absolute size – to address point *ii*).   Some indirect evidence available in the empirical literature, as well as casual inspection of the data, suggest that small countries are likely to be characterised by higher than average per-capita amounts of the natural resource which attracts tourists[10].

In this section we show how, in our dynamic setting, comparative advantage in tourism depends on the size of the natural resource suitable for tourism development *relative* to some measure of the country's size, such as its overall population. In a rather traditional fashion, heterogeneity in relative endowments of natural resources explains the pattern of specialisation obtained in the international marketplace. Clearly, this analysis can be regarded as an explanation of point *ii*) only insofar as the relative endowment of natural resources is shown to be significantly higher in small tourism countries than in the non-tourism countries, especially the larger ones. Our suggestion is that maybe absolute size is just a proxy for the true, economically meaningful variable associated with natural resource endowments. Detailed empirical work is clearly needed to validate this suggestion.

The discussion in section 3.1 makes clear that the heterogeneity in relative endowments should be formalised while controlling for the growth effect associated to the size of the labour force. Therefore, our first assumption is $L_i = 1 \ \forall i$. Second, we assume that a limit exists to the capacity of the tourism sector to absorb the labour force. More precisely, we assume that *a*) the resource suitable for tourism development is an exogenous natural endowment, $\overline{R}$; *b*) this resource is combined in fixed proportion, at zero costs, with labour to produce tourism services. Let us define the fixed quantity of $\overline{R}$ per unit of labour required by the tourism technology. Then the sector's production function is:

$$y_T = \rho h_T L_T \qquad (6)$$

Given this technology, the maximum amount of labour an economy can allocate to the tourism sector ($\overline{L}_T$) is constrained by the natural endowment according to:

$$\overline{L}_T = \overline{R}/\rho \qquad (7)$$

For simplicity, let us choose units so that *p*=1 and therefore $\overline{L}_T = \overline{R}$. With endowments not uniform across countries, a useful heterogeneity may emerge. Countries with a relative large endowment of $\overline{R}$ ($\overline{R} \geq 1$) have the option of allocating their whole labour force to tourism; this option is not available for countries with smaller endowments ($\overline{R} < 1$).

Given that in our framework all countries are "small", this is the closest we can get to the idea that in reality small countries are more likely than greater ones to have a large endowment of the appropriate natural resource relative to the size of the labour force. With this kind of heterogeneity, the dynamics of the system under an autarchic regime gives rise to a resource-based comparative advantage that, together with the result discussed in section 2, offers a unified explanation of the two above-quoted points *i*) and *ii*). To see this, we first recall the determination of the autarchic steady state for a non-constrained representative country. Later we will introduce the constraint and evaluate the consequences on the determination of comparative advantage.

Define *q* as the price of tourism relative to the price of the manufacturing good in autarchy. Pure competition implies that the rate of change of *q* is equal to:

$$\frac{\dot{q}}{q} = \frac{\dot{h}_M}{h_M} - \frac{\dot{h}_T}{h_T} = \lambda_M - (\lambda_M + \lambda_T)L_T \qquad (8)$$

The level of L_T in each period is obtained solving the model for the static momentary equilibrium. In this equilibrium we have[11]:

$$L_T = \left[(\alpha_T / \alpha_M)^{-\sigma} q^{\sigma-1} + 1 \right]^{-1}. \qquad (9)$$

For σ<1 the first derivative of (9) with respect to q is positive, and the second is negative. Using (9) in (8) therefore we see that in this case a stable steady-state value of q exists [such a steady state is unstable if $\sigma > 1$, as in Lucas (1988)]. For our purposes, it is worth studying this equilibrium in greater detail. Equation (8) indicates that the value of L_T corresponding to the stationary value q^* is:

$$L_T^* = \frac{\lambda_M}{\lambda_M + \lambda_T}. \qquad (10)$$

We plot the right hand sides of (9) and (10) in Figure 3.1. Notice that for any given value of q', the difference between L_T^* and $L_T(q')$ yields a measure of the rate of change of q. More precisely,

$$L_T^* - L_T = \frac{\dot{q}}{q(\lambda_M + \lambda_T)}. \qquad (11)$$

As for comparative advantage, if all economies are similar, they will all end up with q*, and no long run pattern of comparative advantage emerges.

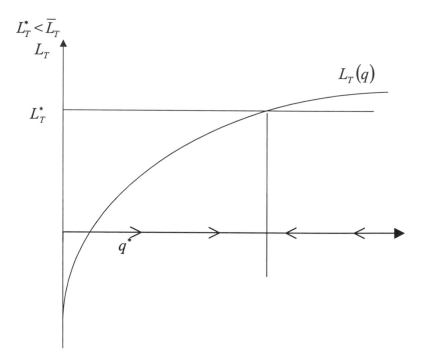

Figure 3.1. Not Binding Natural Resource Endowment Constraint

This is not so if countries are characterised by a sufficient degree of heterogeneity. Assume that resource endowments are such that the constraint $\overline{L}_T < 1$ characterises a subset of countries. Two possibilities now arise. The first is that:

$$\frac{\lambda_M}{\lambda_M + \lambda_T} < \overline{L}_T < 1 . \tag{12}$$

In this case the constraint has no consequences on the determination of comparative advantage. If, instead,

$$\overline{L}_T < \frac{\lambda_M}{\lambda_M + \lambda_T} < 1 \tag{13}$$

then in these economies a stationary value of q does not exist, since q grows at a positive constant rate equal to $\lambda_M - (\lambda_M + \lambda_T)\overline{L}_T$ (see Figure 3.2).

In the long run the countries in this subset produce both manufacturing and tourism goods, with a stable (constrained) allocation of labour. However, such a stable allocation implies an ever-increasing relative price of tourism. The consequence for comparative advantage is straightforward. In the long run, countries in which the resource constraint is not binding end up obtaining the stationary price q^*. Countries where the constraint binds end up with a higher (and increasing) relative price of tourism. Notice that this pattern of comparative advantage is independent of the countries' initial conditions, and that this is so because $\sigma < 1$[12].

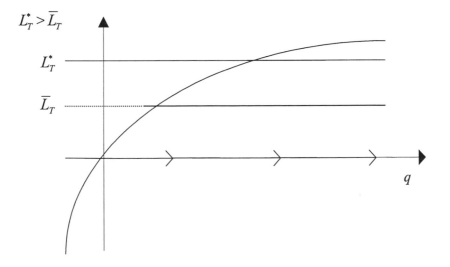

Figure 3.2. Binding Natural Resource Endowment Constraint

When we add up this finding to the outcome described in section 1, we find that, as long as the elasticity of substitution in consumers' preferences is low enough: *a*) countries with large endowments of suitable natural resources relative to the size of their labour force are likely to develop a comparative advantage in tourism; *b*) these countries grow faster than those who specialise in the manufacturing sector.

DISCUSSION AND CONCLUSIONS

Cross-section data on economic growth and tourism specialisation for the decade 1985-95 show that a number of tourism countries grow fast and are small. In this paper we have proposed a joint explanation based on a simple two-sector endogenous growth model. While the model highlights an important reason why tourism specialisation is not harmful for growth, the explanation put forward for the issue concerning size is perhaps more tentative. In this respect, our analysis suggests that what matters for explaining specialisation in tourism is a country's relative endowment of the natural resource, rather than its absolute size. Our suggestion is that maybe absolute size is just a proxy for an economically more meaningful variable. To validate this suggestion, future research should produce detailed evidence on the existence of a negative relation between absolute size and per-capita amounts of natural resources.

More generally, we need further empirical work to assess the robustness of the two rather scanty empirical observations presented in this paper. Among others, two questions are worth underlining. The first concerns the size issue. How robust is the relation between size, tourism specialisation and growth?

Second, the whole analysis developed in this paper assumes that the observed growth rates are a reliable representation of the long-run rates – the only ones we have analysed theoretically. Problems with this assumption are particularly acute in the case of growth based on natural resources. A non-optimal use of the natural resources might initially spur high growth rates that turn out to be unsustainable in the longer run, in a way similar to the one described by the literature on the so-called "Dutch disease". In our model, supplying the tourism service does not reduce the stock of the resource available in the economy. However, were such a reduction to take place, an important consequence would soon emerge: the process might easily induce the country's loss of comparative advantage in tourism, with negative consequences for its long-run growth rate. As a consequence, one key empirical question that remains to be answered is how sustainable the currently observed high growth rates of a number of tourism countries are in the long run.

NOTES

1. World Development Indicators, 1998
2. The first list includes the following countries: China (9.0%), Korea Rep. (7.7%), Thailand (7.6%), Suriname (7.0%), St. Kitts and Nevis (5.9%), Singapore (5.4%), Antigua and Barbuda (5.3%), Chile (5.3%), Indonesia (5.3%), Maldives (5.1%), Mauritius (5.1%), Malta (4.8%), Hong Kong (4.6%), Seychelles (4.5%), Cyprus (4.5%). In brackets the average annual increase of the per capita income from 1985 to 1995.
3. This second list includes Antigua and Barbuda (94.7%), St. Kitts and Nevis (42.3%), Barbados (41.1%), Grenada (27.6%), St. Vincent and the Grenadines (25.6%), Cyprus (23.6%), Jamaica (22.9%), Seychelles (21.7%), Dominica (17.9%), Gambia (10.8%), Mauritius (11.8%), Jordan (11.2%), Singapore (10.8%), Dominican Republic (10.7%), Guyana (10.4%). In brackets the average specialisation degree over the period 1985 to 1995.
4. Using a sample of 200 independent states, Liu and Jenkins (1996) find that in 1990 a significant negative relationship exists between the ratio of tourism receipts to GNP and population size. In the present paper, we adopt this simple definition of country size.
5. This assumption may be justified in terms of the importance of services in tourism and of the fact that, over a long period, productivity growth in services has lagged behind that in manufacturing. For instance, among the OECD economies as a whole, output per person employed grew between 1960 and 1993 by an average of 1.6% *per annum* in services but by 3.7% in manufacturing [Temple (1997)].
6. The joint presence of intersectoral of knowledge generates substantial changes in the results of Lucas's model. In particular, their presence, when combined with that of international spillovers, tends to rule uneven growth out. See Murat and Pigliaru (1998).
7 More detailed results are presented in Lanza A. and F. Pigliaru (1998)
8· On the relation between absolute size of the natural resource endowments and economic growth, see also Sachs and Warner (1995), and Gylfason, Herbersson and Zoega (1997).
9. This scale effect is typical of the class of model to which Lucas's belongs. All the learning-by-doing models of this type are characterised by a scale effect attached to the endowment of the fixed factor of production (labour, in our simple case). In general, normalisation is adopted in these models precisely to get rid of such effect, which many

economists would regard as a rather "counterfactual" one [see Barro and Sala-i-Martin (1995)].

10. For instance, Liu and Jenkins find a strong negative correlation between the log of tourist arrivals per square meter and the log of country size [Liu and Jenkins (1996), p. 112].

11. See Lucas (1988).

12. With $\sigma > 1$, all constrained countries would obtain a comparative advantage in manufacturing, but the comparative advantage of the unconstrained countries would crucially depend on their initial conditions.

REFERENCES

Aghion P. and P. Howitt (1998), *Endogenous Growth Theory*, The MIT Press, Cambridge, MA.

Barro R. and X. Sala-i-Martin (1995), *Economic Growth*, McGraw-Hill, New York.

Candela G. and Cellini R. (1997), "Countries' size, consumers' preferences and specialization in tourism: A note", *Rivista Internazionale di Scienze Economiche e Commerciali*, 44: 451-57.

Grossman G. and Helpman E. (1991), *Innovation and growth in the global economy*, The MIT Press, Cambridge.

Gylfason T., Herbertsson T.T. and Zoega G. (1997), *A mixed blessing: natural resources and economic growth*, Centre for Economic Policy Research, D.P. 1668

Lanza A. (1997), "Is tourism harmful to economic growth", *Statistica*, 3.

Lanza A. and Pigliaru F. (1994), "The tourism sector in the open economy", *Rivista Internazionale di Scienze Economiche e Commerciali*, 41.

Lanza A. and Pigliaru F. (1995), "Specialisation in tourism: the case of a small open economy", in P. Nijkamp and P. Coccossis (eds.), *Sustainable tourism development*, Avebury, Aldershot

Liu Z. and Jenkins C.L. (1996), "Country size and tourism development: a cross-nation analysis", in L. Briguglio et al. (eds.), *Sustainable Tourism in Islands and Small States: Issues and Policies*, Pinter, London, 90-117.

Lucas R. (1988), "On the mechanics of economic development", *Journal of Monetary Economics*, 22: 3-42.

Murat M. and Pigliaru F. (1998), "International trade and uneven growth: a model with intersectoral spillovers of knowledge", *Journal of International Trade and Economic Development*, 7: 221-36.

Sachs J. and Warner A. (1995), *Natural Growth Abundance*, Harvard Institute for International Development , Development Discussion Paper, 517a.

Sinclair M.T. (1998), "Tourism and Economic Development: A Survey", *Journal of Development Studies*, 34: 1-51.

Temple P. (1997), *The Performance of U.K. Manufacturing*, Centre for Economic Forecasting, London Business School, Discussion Paper 14/97.

4. THE STRATEGIC IMPORTANCE OF THE CULTURAL SECTOR FOR SUSTAINABLE URBAN TOURISM

ANTONIO PAOLO RUSSO AND JAN VAN DER BORG*

1. INTRODUCTION: A SUSTAINABILITY APPROACH TO THE IMPACT OF TOURISM ON THE CITIES OF ART

The impact of tourism activities on urban areas has been the subject of a comprehensive amount of research (Briassoulis and Van der Straaten, 1992, UNESCO/ROSTE 1993). Most of these studies focus on the processes of crowding-out caused by the huge pressure of tourism-related activities, relatively little space intensive and able to pay for high rents. Models of urban land use yield the optimal number of visitors that leaves unviolated the carrying capacity of the sub-system of which the city consists, or the optimal mix between categories of visitors characterised by different budgets and mobility patterns (Costa and Canestrelli, 1991). In a dynamic setting they can predict the optimal side-payment associated to a restriction to the visits (Batten, 1991).

However, it is common observation that economic research finds a bottleneck in translating to policy practices. The recognition that demand-side measures are often useless or easily side-stepped led to an increasing interest towards supply-side measures, bound to increase the added value of tourism while at the same time minimising the negative impact on other economic and social functions of the town.

Often, though, even supply policies have proved ineffective, because of lack of control of the local governments over the financial resources to pursue autonomous tourism policies and develop adequate incentives for the local actors to move in accord with the urban strategy.

Hence, the necessity of moving from the analysis of the tourism sector as "isolated" or "opposed" to other sectors to explore the feasible synergies of a sustainable tourism sector on the local economy and society. The studies which explicitly and thoroughly try to evaluate the impact of tourism on the local economy[1] adopt a static approach: a qualitative assessment of the capacity of tourism to positively impact on the pattern of urban development is absent, though comparative static studies and input-output analyses (like Fletcher, 1989) of the local economy are by no means useful to judge upon the potentialities of the system.

The socio-economic impact of tourism in an area is strictly linked with the characteristics of the demand and the organisation of the supply. The demand side of the tourism market is characterised by the place of origin of the visitors, their motivations, and the mode of use of the place of each of the segments. The number of overnight stays, their distribution and the divergence between the region of imposition of costs (extension of the area visited) and the region of benefits (where the visitors spend their budgets) represent good indicators of the economic impact of tourism. In areas where such divergence is large, the tourism pressure becomes unsustainable; common features of such scenario are boosting central rents, a high share of excursionists on the visitors' flow, the start of an unrecoverable urban and fiscal crisis, the drain of resources destined to the maintenance of monuments, and ultimately the crisis of the tourism sector itself, due to congestion and decline in the quality of the product. The structure of the visitors' flow – so characterised – determines how heavy the pressure of the tourist demand on the site is.

The relation of tourism demand with the organisation and the spatial implications of tourism supply are well described by the theory of the life cycle of tourism destinations: the attractivity of the resort is thought to follow a deterministic path. Cities that are able to reach the critical mass in terms of tourist attractivity take off and reach maturity; then, when the costs imposed by tourism activities taking place in the area begin to outweigh the benefits, tourism – if unmanaged – may eventually decline.

Each of those stages is associated to a certain spatial distribution of costs and benefits from tourism, and to a well-defined composition of the visitors' flows (Van der Borg, 1991). The share of day-trippers in the visitors' flow is of particular importance, because excursionists impose a huge amount of costs on the town without contributing but to a very small extent on their financing. Since the importance of their share is associated to the first and the maturity stages of the urban life cycle, a successful management of this segment (in terms of minimisation of their impact and of "fidelisation" – have them sleeping in the city's accommodation rather than commuting) is the key issue. At the same time, the information about this kind of visitor flow is scarce and not systematically collectable. Excursions do not imply an overnight stay in the accommodations; the visit is concentrated in a single day. An art city can be the destination of up to 6 million excursionists every year, as in the case of Venice, where they represent 80% of the overall mass of visitors.

In general, pro-active policies – aimed at ensuring in advance the conditions for the sustainability of each forecoming stage and at minimising the conflicts that emerge in the different stages – are required to maintain a stable path of tourism development.

For our purposes, we will consider how these situations are connected to the performance of the cultural sector of the city. As Thorsby (1994) points out, the economic system and the cultural system can be described as parts of a unique evolutionary model, simultaneous determinants of its sustainability. In places where the tourism demand is low, the two industries act like separate but interfacing entities: culture is the feeder of the (scarce) tourism activities; the act of consumption by the resident population is often what attracts visitors. In this setting, culture remains attached to the values and the environment of a community, and the process of production of culture is strictly linked to the local organisation of society. This explains the strong cultural imprinting of rural communities or of suburban neighbourhoods, in which tourism is seldom allowed to become part of the process. Yet, the danger is that there are insufficient resources to keep those cultures lively and fertile: the critical mass for "profitable" (in a social sense) institutions which hand over and refresh these values cannot be found, and sometimes they die a natural death.

On the other hand, where tourism pressure is high and the profits for the stakeholders of tourism are enormous, the opposite may occur. In those tourist towns there is no incentive for selective marketing or de-marketing: the supply is saturated for most of the year, no matter the quality of the tourism product. Tourism is the dominating feeder of cultural consumption, is fully part of the production process of culture, and the cultural content of the products supplied is the outcome of an optimising behaviour of the tourism operators. It is an input as well as a by-product of tourism, in the sense that it is strongly influenced by the tastes and modes of production of a group of visitors whose set of cultural values of reference is almost infinite, and by factors over which cities have very limited control (Bianchini, 1993). The cultural services supplied are those the tourists ask for, no more no less than a mere representation of the visitors' expectations and cliché images. There is no perceived need to organise a cultural sector: the value of tourism remains in hotels, guided trips and cheap souvenir shops, with churches and museums acting as simply tourist traps. Yet, as we know from the theory of life cycle and observation, this is not a never-ending process; there exists a threshold beyond which – if unmanaged – the attractiveness of a tourist town stagnates and then declines.

The fact is that often – and in medium and small sized art cities this seems to be the rule rather than the exception – tourism operators are not choosing their actions according to the right time and space scale. Not on the right time scale, because their behaviour is inherently myopic, being the case of a multitude of small operators of the traditional type, with an high turnover to other sectors, risk-adverse, and scarcely organised (CISET 1996); not the right space scale because most of them live or operate away from the area of costs. They are not citizens themselves, and they are not aware they are spoiling the golden-eggs chicken[2].

Between the two extremes, of course, there exists an optimal level of tourism demand in a dynamic sense, with a rate of growth equal to that of the other sectors of the economy. In this utopistic setting, the tourist sector has a "natural" impact on the other sectors, the crowding out effects are limited, and a lively and creative cultural sector can grow, of which tourism represents a powerful – but not exclusive – demand basin.

In both the "low" and "high" demand situations, tourism needs to be properly managed: in the former case because if it does not reach the scale to become a self-reproducing industry (the take-off stage of tourism destinations life cycle) its influence on the local economy is limited and generally negative (UNESCO/ROSTE, 1993). In the latter because a "virtuous" development of tourism in relation to the other urban functions must be looked for, whereas the natural trend of market forces and short-sighted behaviour would push the vicious spiral of auto-phagocytation to the extremes and cause the maximum of conflicts for the use of urban space.

In this case, the good management of the cultural sector is a crucial point, because in the case of art cities, culture is what ultimately attracts tourists; and choking the cultural production is a threat for the attractiveness of tourism purposes in general. If, as it is the case in many tourist destinations, there is also an undergoing trend towards tourism mono-functionalism, this means that the whole urban economy could be seriously affected.

The interest we place on the cultural sector is not occasional: it is in the production of culture that medium- size art cities all over the world have a competitive advantage, and – if a broader definition of culture than merely fine arts and performing arts is adopted – it is the cultural industry which offers the most promising developments for urban economies in the new century (Castells and Hall, 1994). Once taken off, the synergetic relationship between tourism and culture allows the local economy to prosper without a significant intervention of the public sector, which can limit its role to steering through the formation of partnerships with the private sector.

The point is: how can this virtuous synergetic development be initiated? What are the elements that guarantee the success of such a strategy? What are the conditions that favour this development, and the nodes that prevent it from arising spontaneously?

To answer these questions, our attention will turn to the cultural sector (section 3), namely the cultural institutions (in a broad sense), the relation and networks between them, the authorities which make the planning for culture, and the actors and technological solutions which can influence the economic outcome of the consumption of culture by tourists and residents. In section 2 we will consider some common features of models of tourism management which do not take into the right account the interrelations between systems in the urban economy. Finally, section 4 will deal with the key concept of "organising capacity" in the cultural sector, and section 5 will assess some cases of planned cultural development according to that framework.

2. MODELS OF URBAN TOURISM MANAGEMENT

From the previous section, it should be clear that a management model for the tourist city which considers the tourist sector as an "isolated" sub-system, with one-way interrelations with the other urban functions, based on a strategy of "solving conflicts" or "minimising impacts", can be profitable to the tourism businesses in the medium term, but may prove unsustainable for the urban economy on the whole. We could even sketch out a paradox of a very successful tourist management strategy which increases the value of tourism business so much that the urban economy gets transformed in a tourism mono-culture. The objectives of such a management model are the following: a) to promote tourism through a sectorial strategy; b) to maximise the profits of the sector; c) to guarantee its continuity; d) to defend its position respect to the other sectors; e) to compete with other tourist destinations.

An alternative model of tourism development is one that establishes and manages synergetic two-way relationships with the other sectors of the urban economy (Figure 4.1); its objective is to maximise the performance of the urban economy as a whole and promote the growth of the strategic sectors. Some characteristics of what we could call the "synergetic model" of tourism management are the following:

- it strives to maximise the impact of tourism on the other sectors of the urban economy;
- it fosters the development of supplier service sector;
- it derives its input from a productive and lively cultural sector;
- it aims at optimising quality rather than maximise quantities;
- its approach is integral and of long-period

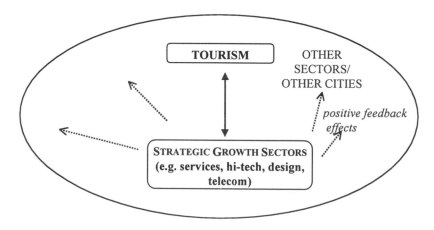

Figure 4.1. A synergetic model of tourism management

In the long term this is a sustainable strategy, because it avoids the risk of a monoculture, while at the same time providing the basis for the growth of other productions. Yet the establishment of synergetic relationships with the strategic sectors of the economy – and in particular with the cultural sector – is by no means an automatic task.

In fact, the endowment with an important cultural and architectural heritage is neither a sufficient condition for an innovative and durable cultural sector to grow, nor a necessary one, as it is demonstrated by the cases of – respectively – many art tourism leader cities and many former "cultural capitals of Europe". The study of different cases of urban tourism development allows us to draw a distinction between a spontaneously exploitable cultural heritage attracting albeit huge tourist flows, and an organised cultural cluster reproducing itself in a virtuous synergy with the other economic functions performed in the urban area.

The latter case is that of cities with a far inferior heritage to count upon and without a strong tradition of tourism destination. Among those cities we can count Antwerp, Glasgow, Copenhagen, Bilbao. These cities were freer to organise a profitable cultural sector, respect to which clustering of different strategic activities was a crucial feature.

Some common elements which are to be found in the way those cities dealt with development policies are the following:

- a general dissatisfaction with the economic performance of the urban area, and in particular a perceived incapacity to attract visitors, residents, and economic activities; this corresponds to a "low growth" position in the curve of the life cycle of a tourist destination (the "take-off" stage in the case of Glasgow, Antwerp and Bilbao, the "stagnation" and possibly the "decline" stage in the case of Copenhagen);
- the transition of the local economies from an industrial or commercial specialisation to an advanced service industry, with the related necessity to re-position the city in the new hierarchy of urban functions and networks and a general need to improve the infrastructures so as to adapt them to the requirements of the new economy;
- the recognition of the potentiality of the cultural sector as a breeding ground for new businesses linked to the hi-tech and hi-touch industries, as well as a means for social development and upgrading of the town's skills and image;
- a scarcity in the endowment and performance of cultural institutions, exposition halls, theatre companies, nightlife entertainment, etc.;
- the widespread agreement on the need for a strategic market planning process as the basic instrument for an urban recovery.

On the opposite front stands the experience of some "European queens" of cultural tourism. In most of those cities, tourism has already reached the maximum threshold of the carrying capacity, and it seems that the risks of entering the "declining stage" of the life cycle are very high.

In those cases, among which are big attractors like Venice, Salzburg,

Florence and Bruges, the cultural character of the visits is still very strong, but two elements are slowly coming to evidence: a) the cultural sector and the single institutions have lost a lot of their attracting power, or the city authorities feel dissatisfied with the performance of those attractions in the overall weight of the tourism sector; b) the cultural production in the town is poor, is getting more and more banal, and the research institutions which should keep alive the cultural tradition of the town are increasingly faced with funding shortages and identity problems.

3. CULTURAL POLICIES IN THE CITY AND THE ORGANISATIONAL ISSUE

Whenever supply policies are effective in smoothing the most negative impacts of mass tourism, they are of little help for general crises of the cultural sector. The cutbacks of local government budgets led to a severe stagnation in the policies of acquisition and conservation in most European countries; at the same time, the increasing fiscal crisis caused by the tourist use of the town threatens to make these situations unbearable even in the case of an augmented financial autonomy of the cultural sector. While it is not the aim of this paper to go deep into the problems affecting the liveliness of the cultural production process, which is more a matter of sociological or anthropological studies, there are some issues that are relevant to the researcher in urban economics.

The most important is an organisational issue. The demand for organisation of the network of the cultural sector emerges particularly when it becomes evident that arts activities should become an integral part of the physical development of the city and region (Perloff, 1979). The different actors of the cultural sector should be included into the decision making process of urban development, and the management of the cultural sector should be unified and homogenised to ensure the maximum productivity of the cultural system, optimising two-way relations with the other urban functions, tourism included. This means, for example, that the commercialisation of the tourist product and that of the cultural product have to be supported by the same structure – be it the technological support or the incoming offices; that the marketing strategy of the destination should be based on the cultural opportunities; that the managers of the tourism industry train themselves and share the same know-how of the managers of cultural institutions.

The interrelations with tourism are no longer mere externalities in production, then, but co-ordination of the processes: in this way the meaning of accessibility changes from accessibility of contents to accessibility of fruition, and the strategic importance of the cultural sector is improved.

As a particular case, the applications of information technologies and telecommunications (ITTs) to the tourism industry as a tool for regulating the flows while at the same time allowing an increase in the added value of the cultural product supplied have been widely studied (Buhalis, 1996). This is an immediate example of how a two-way relation between tourism and the service sector can be established. By sharing the hard networks and integrating the soft apparatus by means of interactive and learning information suppliers, the city offers an opportunity of creative and valuable visits at affordable prices, achieving a set of objectives which fully inscribe in the sustainable approach:

The regulatory objective: remote access to information, the possibility of advanced reservation and booking, the incentives to cultural fruition with the issue of smart cards favouring the access to cultural institutions and events, give the possibility to monitor the incoming flows, respond in real time to peak situations with (dis)incentive policies and special offers for groups or categories of visitors, and promote off-season visits. In the end the ITTs allow internalisation of tourism costs while spreading (locally) its benefits.

The quality enhancement objective: as they enhance accessibility to the cultural supply, increase the informative content of the visit, differentiate the supply, allow the personalisation of itineraries, the new technologies significantly contribute to increase the quality of the product; the same media can be used by citizens, schools, etc, promoting the cultural consumption of the residents and achieving a social objective.

The profitability objective: the convenience of overnight stays is increased, visitors have more time and the consumption of culture is favoured; when the average time of the visits shrinks, the decentralised attractions – important as they might be – are the first to be excluded from the budget spending of a potential visitor[3]. The re-organisation of visitor management through the ITTs gives a chance to the cultural institutions led by dynamic and creative managers to overcome their financial shortages, adding value and informative content to the product supplied.

The interindustrial objective: According to Perloff (1979), "the potential of the arts – and of the cultural sector in a broad sense – stands in the important contribution it can make to the physical development of the city in such a way as to have substantial economic [and social] payoffs, ..., to the enlargement of tourism, the convention trade, and other attractions to visitors; and, in general, to the economic viability of urban communities". The development and application of new technologies to serve the cultural industry has an induced impact on the productive structure of inner cities, favouring the reconversion to the service sector of declining neighbourhoods and stimulating young professionals and well-educated households to move into town, in contrast with the trend towards de-urbanisation which is heavily affecting most medium size cities all through Europe. These dynamics will be investigated further in the following section.

4. URBAN RECOVERY AND ORGANISING CAPACITY IN THE CULTURAL SECTOR

Two concepts will be treated here which allow us to sketch a "virtuous scenario" of the economic development of tourist cities. The first one is that of cluster, and the second is that of organising capacity: we will refer both of them to the organisation of the cultural sector and its relations with the urban economy.

For cluster we intend a concentration of economic activities linked horizontally or vertically (or even diagonally: see Poon, 1993) in the same production process which strongly characterises the local economy. The reference is often to clusters of enterprises or small firm communities, integral elements of the organisational structure of industry known as "flexible specialisation". This kind of industrial structure, described by Piore and Sabel, is particularly fertile in urban settings where the physical nearness represents a "superconductor" of ideas and cultural values; it also makes the development of institutions and their interventions more effective (Van Dijk, 1995). The links between the economic units take the shape of subcontracting relations and customer services, of the circulation of know-how and human capital, or just of the common sharing of a "technological paradigm" in the process of production. The competitive environment may also be harsh, but co-operative inter-firm relationships and collaborations are also intense. Therefore, a cultural cluster could comprehend the cultural institutions such as galleries and theatres, firms providing technological infrastructures for exhibitions (e.g. physical structures, lighting, multimedia supports and complementary services), software developers, data storage structures, tour operators and travel agents, companies providing ticketing services and promotion, educational institutions and schools, public offices in charge of cultural planning and the managers of cultural institutions, etc.

Clusters guarantee the development of a specialisation which can be spent in the regional organisation of production, in a sector in which the central city has a comparative advantage for proximity. At the same time they help to keep the productive base diversified, because the know-how and the "orgware" developed in the sector can be reproduced as a service supporting other sectors: that is, the small software companies and graphic studios working for museums and opera theatres can easily serve the press industry, the research institutions, or even the banking and insurance sector. But it is in the production of new (sub)cultures, new media, advanced and intelligent hi-tech services and hi-touch design products that the clustered organisation appears fundamental, for the flexible and information-intensive nature that characterises these industries.

The organisation of a cluster is often a task of the public sector, because initially the local economy may not possess the critical mass to take off spontaneously. According to van Dijk (1995), there is no standardised policy formula for triggering off the development of industrial districts, but each situation must be treated with the different instruments of the urban management. Of course, a great effort must be made to analyse carefully the potentialities and the weaknesses of the situation, and to identify the actors involved in a strategy of reorganisation of production according to the cluster idea. A good infrastructure and a pervasive training program, for example, seem crucial elements that the government should be able to provide in order to encourage the subsequent steps of cluster formation, as well as promotion of horizontal co-operation and the reinforcement of existing private support. Organic links should be organised between policies on culture and policies on training, education, research and development (Bianchini, 1993).

The situation of art cities is often one of strong one-way vertical links between institutions of the cultural sector and the tourism business, the strong demander of cultural products; but the horizontal links and the feedback flows between cultural institutions and among them and the final buyers (tourists and residents) are poor. Cultural institutions are often managed by the public sector, and depend on public budgets; their revenues are internalised in the central or supra-local layers of government, while the indirect and induced impacts on the urban economy are ambiguous.

In this context, there is little scope for the externalisation of production with diffusion effects; the industrial organisation of cultural production is much more akin to the old "fordist" mass production than the post-industrial flexibility of the successful cities and regions in the new global economy (Morgan, 1992). This organisational pattern does not foster the innovations of product and process, is not germane to a maximisation of the quality of the products, and in the end is not suitable to face the challenges of international competition and a maintenance of the comparative advantages enjoyed by the art city.

A strategy of reorganisation of the cultural sector which guarantees the maximum impact on the local economy and a virtuous relationship with the demand side must spring from the following:

Investigate the present opportunities. A local economy can display "strengths", but to translate them into real "opportunities" there must be a comparative advantage with respect to competing economies or tourist destinations; so, the fact that a city could be an ideal place for the development of a cluster of the kind described here, does not ensure that firms will be willing to invest in the city. Therefore, to succeed in creating a cluster of culture-related activities, the actors of local development will have to promote and improve the location factors for which the city is less endowed respect to other cities, i.e. the quality of life, the environmental quality, the quality of infrastructure, etc.

Developing a vision and an idea of integral product. Only if a common knowledge of the desired development of the local economy is shared by all the actors involved (e.g. a sustainable tourism sector), the possible divergences upon means are minimised, and the decision-making process is initiated in the most effective way. Moreover, a shared belief in the "mission" of the place is the best way to assure continuity to the policy cycle.

Organising the "missing" links. According to Kooiman (1993), to keep up with the complexity of the modern organisation of society, the governance scheme should reproduce the dynamic, complex and diverse character of the decision making process. This requires a great deal of organisational ability and flexibility of the "new style public manager". The missing links in the scheme of the organisation of the cultural sector, namely, horizontal and feedback links, should be organised by stimulating the creation of institutions where ideas circulate, giving incentives to co-operation, promoting customer audits and providing technical support for the creation of integrated data-bases.

Create adequate institutions to support the reorganised system. The old governmental institutions, with fragmented powers on a horizontal and vertical scale, will no longer be adequate to represent the interests and second the dynamics of the local society. A re-design of governmental institutions is needed – of a kind that "steer more than they row" (Osborne and Gabler, 1992) – to overcome the usual territorial fragmentation and lack of co-ordination of the local governments, and to fully take into account the interests of market forces in the development of policies, although it must be clear from the above that they should not be led only by market calculus.

A scheme that represents in the appropriate way the requirements of a policy change in strategic planning has been developed by Van den Berg, Braun and Van der Meer (1997) (see Figure 4.2). These authors define as "organising capacity" the ability to enlist all actors involved, and with their help generate new ideas and develop and implement a policy designed to respond to fundamental changes and create conditions for sustainable development. According to the theoretical framework of organising capacity, the factors which determine the success of modern cities in the complex, dynamic and differentiated socio-economic environment of today require the ability to create strategic networks as a means to replicate that complexity in the governance scheme.

Successful cities are cities which take into account the relationships between the actors of the economy and their different interests, and manage to organise the interaction among them – and between them and the formal administrative structures – in such a way as to fully exploit the potentialities of development of a region or city. The formation of public-public-private partnerships (PPPs) in relation to single projects is one of the bulks of this approach, but it does not exhaust the orgware necessary for an upgraded capacity to govern the processes that take place in our society.

In addition to the quality of the networking and the degree of co-operativeness, other conditions are crucial, such as a strong leadership, which "can come in different disguises" and guarantees an easy start-up as well as the continuity of the process; the development of a common vision and a strategy about the mission of the region, without which governance lacks the wide-range view and the coherence of the different development projects; the political and social support to the development strategy, which implies a strong capacity to communicate the problems and present envisaged solutions – the creation of forums and audits to involve a large number of citizens in the decision making process can prove strategically important; and spatial-economic conditions that justify the pressure to consider opportunities and threads for the urban economy faced with the new challenges.

Some cities in Europe facing major problems relative to the sustainability of the local development explored the possibilities to overcome the situation developing a strategic market planning process.

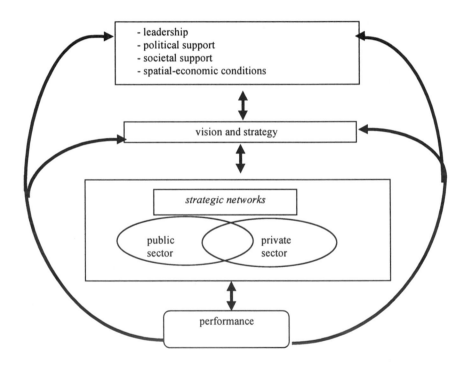

Figure 5.2. Theoretical framework of organising capacity (from Van den Berg, Braun, Van der Meer, 1997)

The dynamic system that these elements contribute to spur up needs inner coherence, but it is on the capacity to handle with single projects and developments that any *a posteriori* assessment on the coherence required can be given.

We can therefore try to evaluate the experiences of some of the tourism cities in the light of this scheme, to find out what the reasons for their success or failure are and what should be done to "correct the route" or engage in a new strategy. Given the subject of this study, it is the capacity to create the conditions for a positive impact of the tourism industry on the urban economy – to be achieved through the promotion of an innovative, lively and attractive cluster of cultural activities and related services – that should be assessed through a closer look to the policies that these cities have implemented (even when they were not directly committed to this specific objective but were generally intended to regulate tourist flows or minimise their impact on the urban economy). The art cities are a main example of an urban context under the threat of unsustainable dynamics triggered by the tourism sector (spatial-economic conditions), and the actors which determine the path of local development are diverse, with diverging objectives and hardly involved in the formal administrative structure. Political and social support to sustainable cultural policies, thus, is not assured; and the experience with networking styles of government may be in its infant stage.

5. SUCCESS AND FAILURE STORIES IN CULTURAL RECOVERY: SOME RELEVANT CASES

These cities shared the recognition of the cultural sector (in a broad sense) as a sector of growth with strong effects on local socio-economic development, because it promotes an economic environment of high added-value small and medium enterprises and it transversally affects the chain of value of different final products for which the demand is high in the "global cities", in particular the tourism product. Telecommunication, press, software, multimedia are the new cultural services which can support innovative and creative institutions to make them profitable. At the same time, the social value of clusters is indisputable, since they contribute to a lively and upgraded living environment, and promote the cultural growth of the resident population, by providing the opportunities to overcome the risk of becoming "divided cities" in the sense described by Fainstein et al. (1992).

These cities – not the traditional tourist destinations – organised a strategy of urban recovery based on cultural planning, which for most of them meant organising a completely new system of cultural attractions (e.g. Glasgow, Bilbao), or re-organising the existing system by re-defining the role of the different institutions in the system and their identity (e.g. Copenhagen, Antwerp, Rotterdam). Most of them were successful: they managed to achieve a reconversion of their productive base with a growth of their producer service sector, and to become attractive for visitors – in particular for cultural tourists – whereas their image had been either that of absolutely unattractive or old-fashioned destinations.

The year in which some of these cities were "cultural capitals" of Europe was at the same time the end point of a planning path and the starting point for new opportunities. The events organised throughout the year brought to the cities an amount of tourists never experienced before. In the end, this process has resulted in a net gain in (permanent) jobs in the service and tourism sector, in an upward structural break in visitors' stays, and in a renewed image of lively cities, culturally stimulating and with an innovative environment. In most cases, these achievements have not ebbed out after a few years. Only in Glasgow the welfare redistribution following the recovery of the inner city aroused a harsh political discussion.

In those cities, the planning of a lively and sustainable cultural sector transcended from a specific calculus of local stakeholders' aspirations; rather, it was organised in such a way as to develop a set of high quality products. An accessible content of culture, or better a cultural product designed around the tastes of an indiscriminate public of visitors, may pose problems of sustainability; an accessible fruition[4] of a cultural product is a very different issue and may be the key to overcome an unsustainable use of the urban resources. New cultural institutions were created, or the old one reorganised, in order to make them part of a system, each of the elements being open to innovations of product (e.g. hi-tech, virtual reality, data bases, complementary services, incentives to production) and of process (e.g. creative financing, different opening hours, integrated management).

Antwerp is the most important city in the Belgian region of Flanders[5], with an urban agglomeration of 1.1 million inhabitants and an economic history of trade, harbour services and fine arts. The reconversion of traditional industrial activities was less striking here than in many other sites in Europe, yet unemployment figures are high and the pressure on jobs is kept up by a huge immigrants flow, which in ancient times was a factor of prosperity and strategic importance. Moreover, the city could count upon the presence of international prominent telematics companies in the region (with a few hi-tech oriented), specialised university education, and progressive telematics policies pursued by local authorities, port authority included. All this, furthermore, in an unorganised and atomised setting.

It is not by chance, then, that the initiator of a recovery policy was the local Chamber of Commerce, and that the strategy heavily relied on two different standpoints: make Antwerp become the "intelligent city" of advanced communication and information processing, and make it become attractive for tourism, leisure and activities related to the cultural industry[6].

The first objective was formalised in a real plan, under the lead of the Antwerp Information Centre. Glass fibre network is bound to link participating public services and companies with an overall reduction of costs and quality of information and transactions and added value to be gained in the future from the processing of such information and the stimulus of network economies by the driving force of hi-tech and hi-touch industries.

The connection with the hospitality and cultural industry followed straightforwardly. The treasures of arts and culture derive from the painting and textile school of the 16th century, but the city cannot be considered a major attraction on the European tourist map, at least as for what regards permanent tourism. Rather, the city could rely on a constant flow of excursionists or short visits from the neighbouring Holland and Germany, with domestic tourism primarily directed to tourism residences and non-hotel accommodation in the province. There were recorded about 3.3 million day trippers every year, corresponding to an expenditure volume of 3,200 million BF. The ratio of these visits to overnight tourism in 1990 was 1.25, but staying visitors contributed with a global turnover of 5,400 BF. Still, the dominant segment was that of business visits. The most visited attractions were not strictly cultural, being the case of the zoo and the cruises in the Flandria region, both regional rather than international in character. A modest number of visitors directed themselves to the very important museums that can be found in town.

The strategy for the promotion of Antwerp as a destination for cultural tourism was inspired by the need to gain new and increased benefits from the existing cultural attractions, linking them closely to the process of cultural production and to the hospitality industry. The quality and diversity of its cultural supply gave Antwerp a strong basis on which to start this policy. The designation of the city as cultural capital of Europe during the year 1993 can be considered the real starting point, also because it inspired drastic changes in the traffic infrastructure. For example, a glass fibre network to a length of over 40 km was laid in record time.

The events connected with the designation of the city as European cultural capital no doubt raised its attraction. The available indicators point to a substantial rise in the number of visitors, with a recorded increase in the number of services given by the tourism board of 180% from 1992 to 1993. What is more interesting, in 1993 – and from then onwards – the city attracted in particular tourists with strong cultural-historical motives, willing to take more than the normal trouble to get information on the cultural product they were visiting. The big mass events attracted an estimated 6 to 7 million visitors, about twice the total number of day trippers estimated for 1990. The question is to what extent the attraction would be maintained or eroded from then on.

It was clear that the existing museum infrastructure and collections could be exploited better by responding more alertly to the international market signals, and by achieving more variety by a more dynamic exhibition strategy. The new programme of the Royal Museum of Fine Arts in November 1993 offered a wider perspective. Close co-operation with the private sector was looked for and found. However, something was missing in the policy of events after 1993: they kept on being oriented to a local and regional public, and planned outside the season, though their potential is still very high. The direct effects on jobs are estimated at around 6,000 man-year for 1993, but it is more in the awareness of policy makers of the opportunities that cultural events offer that the results have to be seen.

So the efforts to consolidate Antwerp's development as a cultural-historical destination continue, despite internal adversities such as the failure of a more general planning policy for regional development which should have brought integrality and spatial considerations into recovery policies. The future of this strategy seems closely linked to two elements: the uniqueness and innovativeness of the local cultural supply, with a production of avanguardistic character given enough scope to be translated into opportunities of fruition; and the application of ITTs to the development and the commercialisation of such product. According to the local authorities for culture, the results were good, with permanent companies and groups stimulated and a fair return in terms of attendance and consensus on quality. For them, this is going to last and Antwerp will be definitively on the map of cultural tourism in a well-developed niche position; moreover, the effects of this placement on the local economy are meaningful. Though the structure created for the year 1993 did not persist, it has left some traces. Informal networks were developed, some producers have found new entrances to sponsors, the number of companies depending on tourism has grown, so that the private sector, too, may be expected to exert pressure and launch initiatives towards a more generous supply of attractions. The massive success of summer events inspired at last the promotion of a summer season. If the suggestions of the tourism policy plan (SPAT) are followed thoroughly in the coming years, the product development will mostly take the form of: stimulating partnerships between museums and private sector; developing a technology-led maritime museum of the Scheldt; developing specific exhibitions and modernising the cinema infrastructures, which has already started; planning the links of the tourist circuit, where a crucial node stands in the organisation of welcoming services and the communication of interactive information.

Is the case of Antwerp to be assessed as a successful implementation of strategic planning for the sustainability of local development? For what regards the role played by the cultural sector in the recovery of the city through a great impulse given to a sustainable tourism industry, the answer is no doubt positive. Once a common vision has been developed in the private sector, the formal institutions to translate it into objectives and concrete policy actions were easily created and also the important counterpart which should provide for technological infrastructures and advanced communication solutions. Some key figures in the local political scene guaranteed a certain coherence in terms of cultural planning, safe-guarding the role of culture from the easy (and blind) temptation of mass tourism production; this turned out to be the crucial factor of success also from a strictly commercial point of view! Political and social support to the strategy has always been assured in a political environment that was awkward and unstable in many other respects; and the continuity of such policy is assured by this success. Paradoxically, the rest of the strategy for urban development was not so successful in many other fields, with the prime obstacle of an administrative structure which has proved inadequate in spatial terms and complexity. From this point of view, tourism and cultural policies have to be considered the vanguard of regional development, even though a lot more has to be done.

The case of Antwerp may be outstanding, in which it represents the clearest example of a strategy explicitly based on the promotion of tourism via a reorganisation of the cultural sector, which achieved noticeable results. In other cases these synergies in development are more tenuous, yet are of great relevance because of the almost hopeless situation at the time before any decision was made. This is the case – among the others – of Glasgow and Bilbao.

The former is the non-plus-ultra of a city in need of recovering policies, or at least this is the image that it has gained world-wide. One of the industrial hearts of 19th century Europe and the "Second City of the Empire", this city had to go through all the painful readjustments to changing global economic circumstances (Pacione, 1995), without completely overcoming its enormous social and economic problems even when modern recovery policies were implemented in the seventies and the local economic environment began to change its focus towards light engineering and services. Deprivation, social fragmentation, housing shortages and high unemployment continued to ravage this city, which was also losing the cultural imprinting embodied in its built capital stock because of the continuous need to accommodate the spatial changing conditions for production: land clearance policies in the inner city began to be attentively selected and some areas of architectonic value were reserved for conservation only well after the post-war period.

However, after many unsuccessful attempts to change the city image and attract new investment and visitors, the card was played of cultural policy and city marketing to dig deep in the dirt of the urban challenge. This resulted in a seven-year campaign ("Glasgow's Miles Better") which succeeded in more than doubling attendance at summer city events; at the same time, the Mayfest Arts Festival was launched and the Burrell Collection opened. This path reached its apogee with the designation of Glasgow as European Capital for Culture in 1990; in the view of commentators, the strategy of image re-building can be considered successful (four out of five visitors found the city "a very interesting and enjoyable place" that year), but the structural causes of the socio-spatial divisions that exist within Glasgow have yet to be attacked. The benefits of this strategy have not reached the most disadvantaged residents, and the drawbacks of such adjustment process have materialised in the boom-and-slump tendency of the housing market in the redeveloped inner-city areas (Jones and Watkins, 1996). Also from a strictly cultural point of view, the top-down approach in reshaping the city's cultural process of production around an year-event was not universally accepted by part of the public opinion, who discarded the thing either as cultural colonialism or a show-off of capitalist power in a city in which a strong value is attached to indigenous working culture and socialist belief. What is new, perhaps, is the ground prepared for private investment and the organisational skills developed thanks to such event; the follow-up is nonetheless considered promising, though important changes in the economic structure of the inner city have to be tested: the "Glasgow Tourism Strategy and Action Plan, 1995-1999" aims to increase the value of tourism in Glasgow by more than 10% by the year 2000, through "innovative cultural development and cultural excellence" (Glasgow City Council, 1997).

Bilbao comes from a similar background, maybe with lesser turmoil behind it than Glasgow. The revitalisation policies pursued, though, were much more effective. The construction of the highly successful Guggenheim Museum can be considered the "cherry" on the cake of a recovery strategy which was sustained by specific organisations, meeting the requirements of the "organising capacity" scheme of governance. And the Guggenheim is not an isolated flagship: the metropolitan area of Bilbao can currently boast seven public theatres, with an average capacity of 1,000, while there were none just 10 years ago. These and other outcomes were achieved in the framework of the project Bilbao Metropoli-30, which also managed to attract two important institutions – the European Software Institute, which functions as a catalyst for related activities, and the European Office for Health and Safety at Work – placing Bilbao on the important map of cities hosting European institutions.

We could refer to Renaissance history to describe another case of tourism development, turning to the twin partner and main competitor of Antwerp on the markets and the arts (Burke, 1993) – Venice – and we will be describing the opposite situation.

Venice is one of the main destinations for cultural tourism in the world. Faced with a non-expanding supply of beds in the centre, and a population decreasing constantly, the tourism pressure on the residents of this city exhibits an exponentially increasing trend (Figure 4.3). The costs of its tourism development are similarly increasing, while a real, integral and widely accepted strategy of tourism management never gained sufficient momentum.

Though Venice can be fully qualified as a city of art, its cultural, historical and monumental heritage is not capable of attracting as many visitors as the city in its generality; that is, people go to Venice because they want to visit the city, walk around and go shopping, but seldom do they ever enter one of its hundreds of cultural institutions. This is a shame, because if it is difficult to make people pay for a visit to Venice (allowing for the internalisation of the benefits from tourism), it is relatively easy to make them pay for single visits to attractions. It is a consequence of the lack of information on the immense historical heritage that can be visited, on one side, and of the length of the visit, on the other, which in the case of excursionists is limited to a few hours. The largest numbers regard the monuments placed in the central St.Mark's area, while important museums – like the Accademia or the Guggenheim Foundation – receive only around one visitor out of the twenty who come into town. A comparison of the available data show how all residential tourists seem to be attracted by a cultural visit, while few of the excursionists do, or they simply do not have enough time to.

Moreover, there is a great heterogeneity in the ownership and the management of those institutions: some of them belong to the state, others to the church, some to the municipality and some are private. The result is that an overall strategy of marketing, information and commercialisation is missing, with no central ticket office, visiting hours and periods overlapping, non co-ordinated investment policies, and so on. Often this heritage is inaccessible, or it is over-congestioned with a consequent reduction in the quality of the visits (Van der Borg and Russo, 1997a). In this situation, the structural economic impact of the cultural attractions on the urban fabric is limited. Indeed, a close scrutiny of the dynamics of employment in the central town of Venice reveals (Van der Borg and Russo, 1997b) that in the areas where the main museums are an overall banalisation of the supply is occurring. At the same time, the chain of value of the cultural product is technologically basic and involves more often regional than local firms. Any of the attempts to enhance the value of the product supplied triggering new activities of R&D faces the main obstacle of the disequilibrated pattern of the visits, which does not favour cultural consumption. As in other observed cases, the establishment of certain areas of cities as "cultural district" produced gentrification, displacement of local residents and facilities, and increased land values and rents (Bianchini, 1993).

The integration of the management and of the promotion of the museum network, even with the application of an integrated open information system linked with the GDSs of the tour operators (with a de-centralised electronic ticketing office) – sharing the same network of the multiservice card system seen in the previous point – is now under study, despite the complexity of the institutional framework. Moreover, a de-centralisation of the cultural supply could greatly benefit from the diversification of access to the town and work as a de-congestioning tool for the central areas.

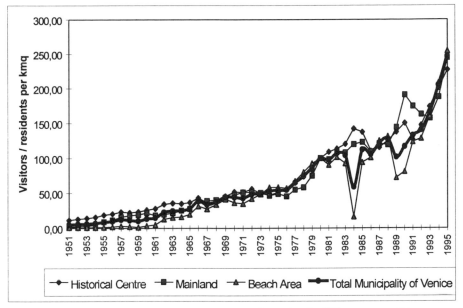

Figure 4.3. Index of tourism pressure on Venice, taken in different areas of the town.
Years 1951-1995

Finally, there is the urgency of a policy of events that may rationalise the temporal pattern of the visits. Only in 1997 an attempt was made to organise a centralised calendar of events, when it became clear among some circles of public opinion and the political forces that control measures on the visitor flows could not solve structurally the existing conflict on the use of the town. Moreover, the monolith of tourism industry was shaken by the recognition that " bad money" chases off the good; in other words, there is an undergoing adverse selection for quality (or aesthetic content) in the supply[7], which could ultimately lead to an overall decline in the attractivity of the resort. The risk of "economic desertification" of the most fancied tourist destination in the world was finally felt by many of the economic agents; at last, the necessity of a timely change in policy gained political support[8], while from the social point of view the city is divided. Yet emergency solutions and last-minute situations always prevail: the Jubilee Year of 2000 will take to Venice an estimated additional flow of 3 to 6 million pilgrims; with the existing structures and facilities, the city runs the risk of being literally submerged, with congestion occurring approximately for 2/3 of the year.

Leadership has been taken by an active and innovative local government, with the mayor re-elected this year with 75% of votes; and spatial-economic conditions for recovery policies, once missing, are now coming back in the form of increasing risk of peripherality respect to a boosting regional economy. Therefore, the next step should be that of setting up the strategic plan to re-shape the city's attraction and to re-build its chain of value of cultural production, in order to meet the global challenges of the next century; in the European map of growing places there is no space for banal processes and backward-looking representations.

The same trends could be observed, at different times and with different magnitudes, in other medium sized art cities in Europe experiencing huge tourist flows (e.g. Oxford, Bruges, Rhodes, Salzburg, etc.), increasingly marginalised in their regional and supra-regional economies and faced with the vicious circle of mass tourism supply. Their impressive heritage and their tradition of tourism hospitality are often the main obstacles to the revitalisation of local economies; therefore, they must make a much greater effort to achieve the conditions for sustainability, and organise an integral strategy in which tourism has an important but not an exclusive role.

6. CONCLUSIONS

This paper intended to provide a general framework for the design and implementation of a tourism development strategy in the city of art; a strategy which responds to the requirements of sustainability, that is the capacity to make the city wealthier without affecting the opportunities of the future generations. Being the case of art cities, our attention could not but focus on the cultural system: it should be reorganised and given a new role in all those situations in which the present pressure is so high that the "normal" process of cultural production becomes inherently prone to the myopic behaviour of local stakeholders.

The reorganisation of the cultural sector must be carefully planned, in such a way as to maximise the opportunities of fruition of a complex and diverse cultural product; to maximise the synergetic effects on other industries; and to maintain the production of culture a lively and creative process.

In this way tourism becomes inherently sustainable, as it provides an opportunity to pay for the value that is generated in the cultural industry; whereas, if ungoverned, the tourist use of the town costs more than it pays. This reorganisation, based on clustering of different activities and the creation of network economies in which tourism, not exclusively, represents the stage of consumption, requires a great deal of capacity by the different actors involved – who have diverse and sometimes conflicting interests. It entails selecting a market, giving a preference to high quality, linking closely the production process of the cultural product to the new innovative productions, decentralising the cultural heritage, setting an integrated approach to the planning of the cultural production.

The theoretical scheme of "organising capacity", applied to the cultural sector, can help to identify and systemise the different elements that contribute to the effectiveness of a policy change: leadership, spatial-economic considerations, political and social support, the quality of networking, and most of all the common development of a vision and an image of the desired development.

Once these conditions are satisfied, the impact of tourism in the cities of art should not be judged only "statically" from the direct, indirect and induced multiplicative effects on the local economy, but – in a dynamic sense – from the amount of new businesses and from the soundness of the economic structure it is capable of fostering through the cultural sector. This will depend on the quality of the links between the sectors and on the overall capacity of attracting investments and organising inter-industrial spillovers.

While some descriptions have been given of how this kind of planning has been approached in different situations and with radically different outcomes, further research has to be done to improve the quantitative analysis of the dynamics of urban change that take place in the cultural city. For example, we believe that the real estate markets display similar patterns in towns that implement effective cultural policies; this hypothesis should be tested by empirical research with data collected on a highly disaggregated level. Moreover, the design of the links between the cultural institutions and the local tourism policy should be accurately defined on a "micro" scale and deserves attention in the field of the management science.

NOTES

*   We would like to thank Leo Van den Berg, Guido De Brabander and Arjen van Klink for their suggestions and constructive criticism.

1.   See for example National Endowment for the Arts (1987); this study contains an evaluation of inter-industrial relations in cultural centres. In the same framework, see Yzewyn and De Brabander (1993).

2.   According to Tronchin (1996), more than 40% of the hotels in the centre of Venice is

owned by people living out of town or firms registered elsewhere.

3. It is the case, for example, of Gallerie dell'Accademia in Venice: due to its peripheral location respect to the tourist route leading to S.Mark's square, this renowned museum of ancient art is only visited by 1 out of 20 of the visitors coming to the town each day.

4. In the declaration of the Alderman for Culture of the City of Antwerp, Mr. Antonisch, we can read: "Antwerp 93 made a conscious choice in favour of art, with a strong emphasis on contemporary art, and not culture in the broad sense of the word .... Art today serves as an antidote to the pressure of work and as an opportunity or pretext for tourism. Art creates jobs ... Accessibility of arts can only mean that everyone who wants to, must be in a position to take part in art as they see fit. The democratic aspects of art have nothing to do with art itself ... Art does not exist without a public; (but) art becomes élitist if those who want access to it are prevented for social or financial reasons".

5. The exposition of the case of Antwerp is primarily based on De Brabander, Chapter 3 of Van den Berg, Van der Borg, Van der Meer, 1995.

6. Other partial recovery policies were implemented to solve the main social problems and could rely on EU funding programs; still an eye to the market dynamics was not missing also in that prospect, since it was carried on, for example, through the economic revitalisation of declining neighbourhoods with the stimulus to small commerce and the creation of new enterprises.

7. A sound economic foundation of these processes is given in Mossetto, 1993.

8. In the campaign of 1997 for the election of the municipal council, all of the parties but one proposed the introduction of a Venice Card as a means to manage the visitor flow.

REFERENCES

Andersen H., Mathiessen C. (1994*), Urban strategies: mega events, A Copenhagen perspective*, Abhandlungen - Anthropogeographie, Copenhagen.

Batten D.F. (1990), *Venice as a 'Theseum' city: the economic management of a complex cultural good*, AIMAC.

Bianchini F. (1993), "Culture, conflicts and cities: issues and prospects for the 1990s", in Franco Bianchini and Michael Parkinson (eds.), *Cultural policy and urban regeneration: the West European experience*, Manchester University Press, Manchester.

Briassoulis H., van der Straaten J. (eds.) (1992*), Tourism and the environment: regional, economic and policy issues*, Kluwer, Dordrecht.

Buhalis D. (1996), "Information and Telecommunication Technologies as a Strategic Tool for Tourism Enhancement at Destination Regions", in Schertler W., Schmidt B., Tjoa A.M. and Werthner H. (eds.), *Information and Communication Technologies in Tourism*, Springer Verlag, New York

Burke P. (1993), *Antwerp: a metropolis in comparative perspective*, Martial & Snoeck,

Antwerp.

Castells M., Hall P. (1994), *Technolopes of the world*, Routledge, London.

CISET (1996), *"Applicazione delle telecomunicazioni e della telematica per la gestione ottimale dei flussi di informazione nel settore turistico: ricerca commissionata dal Centro Studi Telecom Italia S.Salvador"*, Working paper CISET

Costa P., Canestrelli E. (1991), "Tourist carrying capacity: a fuzzy approach", *Annals of tourism research*, Vol. 18.

Fainstein S.S., Gordon I. and Harloe M. (eds.) (1992), *Divided cities*, Blackwell, Oxford.

Fletcher J. (1989), "Input-output analysis and tourism impact studies", *Annals of tourism research*, Vol. 16.

Glasgow City Council (1997), *Glasgow Economic Monitor*, Glasgow.

Jones C., Watkins C. (1996), "Urban regeneration and sustainable markets", *Urban studies*, 33, (7).

Kirchberg V. (1992), "Arts sponsorship and the state of the city: the impact of local socio-economic conditions on corporate arts support", *Journal of Cultural Economics*, 19, (4).

Kooiman J. (1993), *Modern governance*, SAGE, London.

Mabry M.C., Mabry B.D. (1989), "The city as a museum: economic maximizing behaviour in Florence, Italy" in D. V. Shaw, W. S. Hendon, V. L. Owen (eds.), *Cultural economics 88: an American perspective*, Ottawa, Canada.

Morgan K. (1992), "Innovating by networking: new models of corporate and regional development", in M. Dunford and G. Dafkalas (eds.), *Cities and Regions in the new Europe*, Belhaven Press.

Mossetto G. (1993), *Aesthetics and economics*, Kluwer, Dordrecht.

National Endowment for the Arts (1987), *Economic impact of the arts and cultural institutions: case studies in Columbus, Minneapolis/St. Paul, St. Louis, Salt Lake City, San Antonio, Springfield.*

Osborne D., Gaebler T. (1992), *Reinventing the government*, Reading, MA.

Pacione M. (1995), *Glasgow: the socio-spatial development of the city*, Belhaven World Cities Series, Wiley, Chichester.

Perloff H. (1979), "Using the arts to improve life in the city", *Journal of Cultural Economics*, 3, (2).

Poon A. (1993), *Technologies and Competitive Strategies*, CAB International, London.

Severens M.H.H. (1995), *Museumpark Rotterdam: ontstaat er een meerwaarde door het clusteren van culturele instellingen?*, Erasmus Universiteit Rotterdam, The Netherlands.

Throsby D. (1994), *Linking culture and development models: towards a workable concept of culturally sustainable development*, paper prepared for World Commission on Culture and Development, UNESCO.

Throsby D. (1995), *Making it happen: the pros and cons of regulation in urban heritage conservation*, paper.

Tronchin G. (1996), *L'impatto economico del turismo a Venezia*, Degree Thesis.

UNESCO/ROSTE (1993), *Proceedings of the international seminar: "Alternative tourism routes in the cities of art"*, 24-25 June, Venice, Italy, ed. by Dr. J. van der Borg.

UNESCO/ROSTE (1993), *Tourism and cities of art*, Technical Report n. 20.

Van den Berg L., Braun E., Van der Meer J. (1997), *Metropolitan organising capacity: experiences with organising major projects in European Cities*, Ashgate, Aldershot.

Van den Berg L., Van der Borg J., Van der Meer J. (1995), *Urban Tourism*, Avebury, Aldershot.

Van der Borg J., Russo A. (1997), *Un sistema di indicatori per lo sviluppo turistico sostenibile*, Working Paper FEEM n. 05/97.

Van der Borg J., Russo A. (1997), *Lo sviluppo turistico di Venezia: analisi territoriale e scenari di sostenibilità*, Working Paper FEEM n. 06/97.

Van der Borg J. (1991), *Tourism and urban development*, Thesis Publishers, Amsterdam.

Van Dijk M.P. (1995), "Flexible specialisation, the new competition and industrial districts", *Small Business Economics*, Vol. 7.

Yzewyn D. and De Brabander G. (1993), "The Economic Impact of Tourism and Recreation in the Province of Antwerp, Belgium", in *Perspectives on Tourism Policy*, Mansell, London.

5. THE MANAGEMENT OF CULTURAL GOODS: SUSTAINABLE AND UNSUSTAINABLE DEVELOPMENT OF ART CITIES

GIANFRANCO MOSSETTO

In Italy, out of more than 8,000 comuni, or municipalities, more than 50% are classified, presumed or self-proclaimed, cities of art. They may be as large as Naples, as small as Soana; they may have been transformed by the urbanisation of the 1950's like Syracuse or remained unchanged for 500 years like Pienza; they may be the privileged destination of an exasperating tourism such as Venice, or abandoned to their crumbling splendour like Noto – whatever, they still have a say in the matter of solutions to their generalised decadence and the ever-more pressing problem of city management.

It is necessary to bear this multiform and fragmented situation in mind when analysing the management problems of these cities, and the crisis that has invested them when discussing possible solutions.

1. THE LINK BETWEEN ART AND CITY ECONOMY

The general term "cities of art" has little meaning if we do not specify the relationship that exists among the culture, economic development and tourist flow which characterise each of these cities. Each "city" has its own model which brings together its economy and culture (and therefore also "its art", which is the historical and tangible expression of culture). The changes in this model have provoked radical modifications in the destiny of the city over the course of time. It qualifies, both positively and negatively, its situation and its "tourism" and it conditions its future possibilities.

This model must be clearly understood if we are to check the feasibility of an intervention conceived for the resolution of current problems. It also helps us to judge more accurately the effects of public or state intervention which aims at the preservation, decongestion and development of the city.

2. THE BASIC HISTORIC MODEL

The history of art has demonstrated that, in extremely simplified terms, there are two ways in which the cities of art developed and affirmed themselves as centres attracting vast streams of travellers. The first is the Chartres model; the second that of Santiago de Compostela. In the former case, the city developed where there was a flourishing market, where the flow of people (merchants and their clients) had been historically activated for economic rather than cultural-religious reasons. Art (the cathedral) is a product of economic development (donations made by merchants) which in its turn produces a further flow of people (pilgrims) who are therefore attracted for new religious and cultural reasons. This consequently determines their wealth. In the second case, the city itself develops around a sanctuary. The flow of people (pilgrims) is initially activated for religious and cultural reasons. These reasons (pilgrims' donations and consumption) are then also the primary source of the wealth of the city.

These two situations also allow us to understand how the streams of people can assume similar directions and have similar effects even though the fundamental reasons are different. It can therefore also help us to understand how talking about "tourism" may be just as misleading (bearing in mind the proposals made at the end of Section 1, above) as talking about "cities of art" tout court.

3. CLASSIFYING THE MODELS

Following this line of argument and broadening the analysis of the large statistical chapter on Italian "cities of art", we may hazard a definition of some aggregates of models on which to base the elementary reports singled out above. For each aggregate we have supplied the specific attributes of the culture economy relationship and the qualification of the "tourist" flows in terms of their effective motivations.

3.1. The economy dependent models

The first set is that of the "economy dependent" models, for which, culture, art and their development, and their decadence, depend on the economic trend of the city. This dependence can be of two types, which constitute two subgroups to the main model.

A1.

In the first, culture is the consequence of the development of the city economy, in the sense that culture and art are the final result of a consumer process and therefore of a demand which finds its origins in the economic development of the city. The more the city grows, the more it consumes; the more it consumes, the greater the demand for culture and art. The flow of people is activated by the attraction exerted by the city in its entirety as a centre of economic growth, which is also cultural growth.

People do not go to New York mainly to visit the Museum of Modern Art, but if they are in New York they are likely to visit it. Similarly, people do not go to Milan to see the Brera Art Gallery, but if they are in Milan they are likely to attempt a visit there.

Thus, New York merchants in contemporary art or London merchants in classical art are the most important in the world not so much because New York or London are markets which specialise in art, but because these cities are a concentration of potential clients and above all because they are "markets" (of raw materials, finance, services, etc.) of world wide importance.

A2.

In the second subset there are cities for which culture is, or has been, not a consequence but an instrument of the development of their economy. In this case, investment and consumption in the fields of culture and art are intended as the means by which the economic growth of the city is accelerated, just like any other intermediate investment or consumption. Historically, in Venice investments in art and culture (buildings, churches, paintings, public monuments, celebrations and public events) were brought about through a conscious and consolidated decision to use part of the new wealth as a manifestation of the economic power and success of the city[1]. Venice, that is, behaved no differently from any large modern company that uses a part of its profits in advertising and developing its public image, so as to further increase its profits and turnover. The flow of people towards the city is also boosted through an application of these policies.

3.2. The culture dependent models

The second aggregate of models is that of the so called "culture dependent", which sees, differently from the first, the economy itself depending on culture: in this sense, the former is "founded" on the latter. This can come about in two ways, which are also two subgroups of the main model, and which, in the culture-economy relationship, imply:

B1.

Beneficial dependency. In this case, culture is an endogenous reproducible productive factor which is considered to be fundamental for the growth of the city's economy. In this sense, in Weberian terms, this might well be defined as a "Calvinist model". Cultural reproduction and development are not a means of economic growth but a necessary environmental condition. Florence in the 14th and 15th centuries is an excellent example of this model, along with contemporary Paris. In pre-Medici and Medici Florence (up to the period of Lorenzo the Magnificent), growth would have been impossible without the mercantile culture of the Florentines, which was dedicated much more to private virtue rather than to public magnificence[2].

This was a culture which expressed itself through buildings, private chapels in churches, hospitals and defence constructions, all conceived and destined for the glorification of the individual rather than the community. This was originally more a productive rather than a celebrational culture which, even when it was transformed into the official expression of the Principality, never lost sight of its lay and artisan, and therefore scientific, origins[3]. Lorenzo the Magnificent himself is an example of this, with his capacity not only to be a patron-commissioner but also an artist and a planner at the same time, just as before him Cosimo as well had been an indefatigable planner[4]. The case of Renaissance Genoa is not very different. Here, the construction of the famous Strada Nuova was both a manifestation of the entrepreneurial success of the great Genoan families and a wise example of the first substantial real estate investment of the city.

Paris, despite following the same logical canons, is different. It presents itself as the product of a centralised administrative culture, which is the real, constant element in its development, even when the exogenous thrust of the great celebrational investments of the monarchy and the empire abated. In these cases, "tourism" has both cultural and non-cultural motivations. The city is an active producer of culture and the "tourists" are often a part of this production, just as Manzoni was when he "cleansed his clothes in the Arno" or as the American writers were in Paris between the wars.

B2.

At the opposite end there is the model of harmful dependence. In this case, the economy of the city is again based on culture, but only in the sense that culture constitutes an "input" (but not an "output" any longer) which is consumed in the city's productive process without ever being reproduced.

The city is culturally "dead" in the sense that its economy will function only as long as the culture and art produced in the past will last, as in the case of contemporary Venice. In this example, no new culture is produced. "Tourism" may be either cultural or non-cultural, but it never participates in the production of culture, only in its destruction.

3.3. The residual models

There are also residual models in which culture is precisely a residual product, the marginal result of a process which is mainly extraneous to it. Culture is no longer the effect of economic growth, nor is it its instrument or one of its "inputs". Rather, it is a secondary "by-product" which ends up acquiring its own specific relevance through historic "accumulation".

This happens either:

C1.
Because culture and art originate "residually" for political and not cultural reasons. The Cardinals of 16th century Papal Rome had buildings constructed for themselves in order to manifest the rites of their own power[5]. Contemporary Washington is decked out with museums and cultural institutions (the Smithsonian Institute, for example) which have no roots in the economic structure of the city, but rather in its administrative role as a Federal capital. Leopold II of Lorraine attempted to invite artists and intellectuals to Florence in the 17th century in order to reclaim the role of capital city which Florence was no longer able to maintain spontaneously because of the political incapacity of the later Medici[6]. In these cases, culture is a "residual" product or "by-product" in the Paretian sense of the term, as it constitutes the historical remains of a former ruling "élite". Culture is imported rather than produced in the city. The culture and art of the city, and thus its "tourism", depend on its capacity to maintain and/or develop its political and administrative role.

C2.
Culture is "residual" in the sense that it is what remains when the city has been abandoned by its other historical roles. The city thus becomes a

museum, something which continues to exist despite its own death, as is the case of Williamsburg, Pienza and Rottenburg. Culture and art are the specific cause of "tourism", which is per se specialised.

4. THE MODELS AND THEIR POLICY-MAKING IMPLICATIONS

All of this has important consequences in a diagnosis of the evils deriving from the use or growth of the city and in the prescription of possible cures.

Models $A1$ and $A2$ (the "economic dependent" models) and $B1$ (the "beneficial culture dependent" model) are all characterised by a virtuous cycle which unites economic growth, investment in art and culture and their external effects. Economic growth stimulates, albeit for different reasons in the three different models, investments in culture, which give rise to a growth in the external economies which are available in the city. This gives rise to further expansion in demand and therefore of the city's income.

In this model, which is per se potentially explosive, the limit to expansion is set by congestion. Economic growth, in fact, also generates external dis-economies, which can be expressed as costs deriving from the concentration of the population (and relative flows) following the attraction exerted by the urban pole. If these costs grow more rapidly than the benefits deriving from the investment of a part of the increased urban income in goods with positive externalities, there may be the effect of a sharp congestion: the Black Plague, the epidemics of the 16th century, the slums and widespread criminality of the large western cities of 20th century.

The process of development of the positive externalities is different in the three models because the exogenous variable from which demand and therefore the production of culture depend in these models is also different.

The given models suggest that in the case of "culture as consequence" of the economy ($A1$), demand for culture is dependent on the income and, in the final analysis, on the aggregate demand. In that of "culture as instrument" of the economy ($A2$), demand for culture depends on the aggregate profit of the companies and its specific re-investment in culture, and in that of culture as reproducible "productive factor" ($B1$), by overall supply (or production) itself as well as the productivity of culture. In fact, as Schumpeter would have it, the "animal spirits" of this model are stronger (and therefore their effects all the greater) the more important the works (both mercantile and industrial) of those who are put into action by the model are.

The relationship, in terms of external benefits, between the economy and the cultural endowment of the city, may therefore be hypothesised, ceteris paribus, as the modification of the trends, respectively, of the curve of aggregate demand (greater demand for equal income), of the curve of cultural investment (greater investment for equal profits), and finally, of the curve of production (greater income for equal cultural endowment).

The model of the "cultural dependent" economy of the "harmful" type (*B2*), contrary to the preceding, is characterised by a vicious cycle. The more the city's demand increases, the more the productive art and culture factor is consumed and, as it is not reproduced (and it is often non-reproducible), may even be destroyed. In this case, the hypothesis is that the external dis-economies caused by the consumption of past culture are vertically combined, as "public evils", with external congestional dis-economies causing an implosive relationship between aggregate profits and artistic endowment (the rate of investment in culture is lower than the rate of consumption).

The "residual" models (*C1* and *C2*), finally, can be considered "neutral" as far as the well-being deriving from the tourist consumption of the city is concerned, because in the first model (culture as "by-product of political power") one of the two following hypotheses are realised:

- the city maintains its role as "mirror reflection" of the ruling class (its role as "capital"), and then the model assumes the aspect of those with a virtuous cycle (principally *A2* or *B1*, that is culture as instrument or productive factor in growth),
- the opposite occurs, i.e. the city loses its role as "capital" (Naples in the 20th century, or Florence itself in the 18th century), and it might then begin to move towards the vicious cycle model.

In the second "residual" scheme (that of the "museum" city), this leads to two possibilities:

- stagnation (or degradation), if the trade-off between cultural and non-cultural use of the city is not interpreted carefully (this might cause overcrowding or a fall in investments for conservation);

- stability, if specialisation allows for a virtuous maintenance of the effect on the economy of the use of the good.

The models are susceptible of a simplified formal representation in the following figures, where an orthogonal diagram with four quadrants, allows us to link the characteristic aggregate variables, and that is, respectively: the cultural endowment (or artistic heritage) of the city (w); the total aggregate income or demand (x) (which, following a simplified assumption of a balanced economy, is also equal to the total aggregate supply); the tourist flow, or the number of visitors (q) in a conventional temporal unit (one year, for example); and the total profits which we have conventionally assumed to be net of the external negative effects deriving from congestion (which, for the sake of simplicity, we have considered to be internalisable).

Quadrant I contains a representation of the art-income production function along with its relative constraint (or production limit). If we hypothesise decreasing efficiency for the productive factor culture (w), i.e. that Baumol's hypothesis is acceptable, the production function is increasing but concave towards the X-axis. The second quadrant contains an expression of the demand function, measured in terms of the flow of visitors in the period of time taken into consideration. If we hypothesise a polarising effect of the city and/or that art is a superior good, the demand for the city increases more than proportionally in comparison with the income. This function is therefore also increasing and concave towards the X-axis.

In the third quadrant, this function is compared with the aggregate net profit deriving from its satisfaction. As it pays the penalty for the increasing congestion deriving from the more than proportional growth of visitors compared with income, the profit function will also be increasing but concave in respect of the X-axis. It may also happen that it becomes asymptotic in respect of a limit of maximum profit (congestion limit) and then decrease with a further increase in the number of visitors.

The fourth quadrant, finally, is represented by the relationship between the net aggregate profit of the system and culture endowment, which is expressed by the function of investment (or dis-investment) in the art of the city. If there are no exogenous interventions, its curve defines the rate at which the aggregate profit is re-invested in cultural goods.

In the "virtuous" models (*A1*, *A2* and *B1*) illustrated above, it allows for the net increases in the artistic endowment of the city and therefore dilations of the production limit (see Figure 5.1).

In the "vicious" model (*B2*) or in the "neutral" models (*C1* and *C2*), it may, depending on the case, lead to increases or decreases (see Figure 5.2).

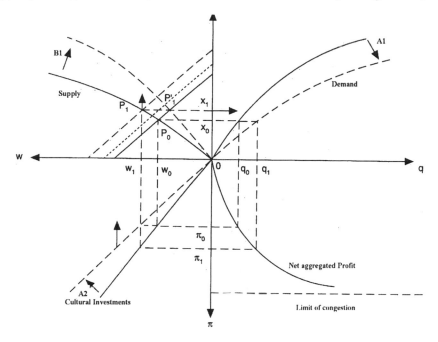

*Figure 5.1.*Virtuos Models

Let P_0 (W_0, X_0) be the point of intersection between the function of demand and the limit of production at which the city is assumed to be in a state of initial equilibrium. If this condition is, hypothetically, stable, the model would then determine to be maintained at W_0 and, therefore, the production at X_0.

Let us now examine the implications of the "virtuous" models compared to this hypothetical situation of initial stability. In model *A1*, ceteris paribus, the position of the consumption (demand) function is shifted downwards towards the right in its quadrant (II). In fact, the economy, that is, the existence of aggregate demand (X), develops per se positive external effects which result in a further growth in demand (that is, of visitors at q_1), and therefore of net profits (π_1) and of artistic endowment (W_1) and the production limit and, again, of income (X_1).

In model *A2*, it is the function of investment which is shifted upwards to the left in its quadrant (IV) as it is the aggregate profit which develops an external effect as it has been re-invested in art and culture to a greater degree than in the other models, instrumentally to the growth of the city.

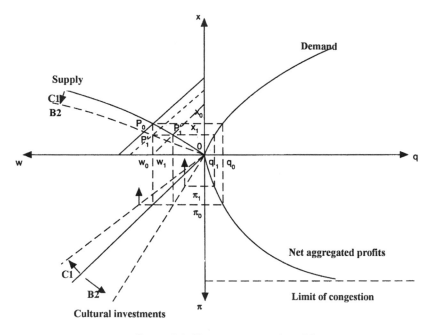

Figure 5.2. Vicious or neutral models

The result is naturally analogous to the preceding one. W is expanded to W_1, and the production limit is consequently expanded and therefore income (X_1) and so on.

In model *B1*, finally, it is the function of supply which moves upwards to the right of quadrant I, as it is the cultural endowment per se which develops a specific external effect. Ceteris paribus, the productivity of the culture is no longer decreasing but probably increasing, as it is representative of technological know-how and the entrepreneurial factor and not simply of past artistic-monumental accumulation. The new limit goes through P'_1, which allows for greater income X_1, and so on.

In all the three variants of the virtuous model, the upper limit to the "explosion" of the economy (or of the expansion of the limit of production) is the limit of congestion, beyond which net aggregate profit not only does not increase but may even diminish (see Figure 5.3, quadrant III).

In the other models (Figure 5.2), the shifting of the initial point of stability P_0 (W_0, X_0) is due, ceteris paribus, either to a change in the inclination of the cultural investment function, or to a decrease in supply following a slump in the productivity of the artistic endowment.

Model *B2* implies a veritable "implosive" economic trend in that the greater inclination of the curve of investment in respect of the "path of stability" accentuating it, always involves a final artistic endowment which is inferior to the initial one ($W_1 < W_0$) and an inferior income ($X_1 < X_0$). The city dis-invests in art and destroys culture, thus decaying. The decadence can also be provoked or simplified by a drop of productivity.

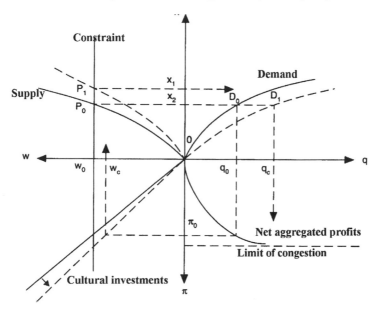

Figure 5.3. Museum city models

In model *C1* the same thing or the direct opposite can occur – in this sense the model is "neutral". If the city loses its political-administrative role, the function of investment shifts to the right, the model becomes that of situation *B2* and the city decays. The same thing occurs if the function of supply is shifted downward and to the left in quadrant I, and that is if it is the productivity of the cultural endowment which diminishes following the loss of the city's role, or provoking it. If the city maintains its role, however, there is shift onto a cultural investment curve which is less inclined than the one that guarantees the "path of stability". This causes an expansion in the artistic endowment and the economy, analogously to that of the virtuous models.

In the case of *C2* (Figure 5.3), the city can either remain in a stable position or expand or shift onto a less favourable curve of the *B2* type, thus decaying. However, contrary to model *C1* the limit of production is parallel to the ordinates as the transformation of the city into a "museum" has by definition already taken place and therefore W_0 is valid for all X (artistic endowment is fixed and cannot be expanded).

In respect of the position of stability P_0, all expansion of the supply and demand towards points of the P_1, D_1 type implies a deterioration of the relationship between profit and cultural investment. In other words, as it is only the productivity of the cultural endowment which can increase, but not its absolute value in real terms, the expansion of the system will only increase the volume of total net profits. When the system is specialised, its expansion is maximal. If we assume that this is the case for initial stability, by definition there is nothing to be gained from a shifting of the stability towards another position. W_0 is therefore also always an endowment of equilibrium which maximises X and π. If, however, the situation of stability does not bring about a complete specialisation, possible shifts of the demand or supply in relation to external effects can cause overbooking (shifting consumers from q_0 to points of the q_c type) (Figure 5.3), or complete destruction of the artistic endowment ($W_0 > W_c$).

5. WHY SUBSIDIZATION OF AN ART CITY COULD BE CRITICAL

These models are useful, as we have already said, not only in analysing the nature and origin of the growth and decadence of the cities of art, but also in studying the effect of possible intervention policies aimed at finding a remedy for the cities' problems. To be honest, historically Italy has always had only one "policy" rather than a series of "policies", i.e. subsidies.

Faced with the economic decadence of the cities of art, their relative depopulation, the degradation caused by an increasingly mono-cultural use of the city by tourists and the gradual erosion of the quality of the cities' artistic heritage, the answer given by the public sector has been that of subsidising with state funds businesses, residential buildings, tourism, restoration, the administration of cultural structures and their production.

These funds are mainly given to private enterprise or local bodies, which tend to promote above all conservational and restoration work on artistic monuments and similar related objects (painting and sculpture). A small part of these subsidies is given over for the running of museums and historical libraries, and an even smaller part to "cultural activities", i.e. the production and re-production of culture. The destination of these subsidies is normally carefully defined well in advance (on the basis of an institutive norm regarding the apportioning of funds) in reference to the nature of the specific works to be financed and the mechanisms by which the money can be apportioned. Local administrative discretionality is limited at most to the establishing of the quality standards which must be met in the conservation projects and the artistic product being subsidised.

This is the sense of establishing regulations for the restoration itself and the duty to see that these regulations are met, and also the choice of the products which may be "generated" via subsidisation (musical, opera, theatrical, exhibitions, etc.). The geographical allotment of the subsidies is mainly determined by the geographical distribution of the works to be preserved. The allotment is therefore "historical" in nature. The real (and only) criterion for the allotment of subsidies (in the best of cases) is the physical decomposition of existing works (or monuments), or the historical sum spent (for subsidies given for the administration of these works).

In general, as for all subsidies which are governed by strict rules regarding destination, these resources have an opportunity-cost (or a shadow-price) which is extremely low for those who have to decide on their effective use. No "fiscal effort" is required at a local level in order to obtain these resources. In principle, there are therefore very few innate safe-guard mechanisms (like the one in play between voters and the people they elect) as far as the use of these resources is concerned. This is coupled with the "prevalence" of a "welfare" mentality which invites those elected at a local level to consider (perhaps quite rightly) those elected to be potential "free riders", so better a subsidy not requested or desired than nothing at all.

The amount of money subsidised therefore sometimes runs the risk of being excessive when compared to the effective requirements of the object subsidised. The same can be said for the tributary instruments instituted with the aim of guiding private initiative (mainly formalising the tributary aspects for the costs involved in restoration work and some forms of donation, included under the heading "tax deductible" in income tax declaration forms both for individuals and for companies). These are, in fact, "tax deductible" costs which are to all extent and purposes analogous to subsidies.

It follows, above all, that public action can be represented in simplified terms in our models through a reduction of the price level at which demand is satisfied (and that is with a downward shift towards the right in the curve in quadrant II); or through an increase in the quantity offered for equal input (quadrant I), or through an increase in cultural investment for equal profits (quadrant IV). Similar interventions have different effects depending on the level of use that the artistic endowment has in the specific situation to which we are referring.

In fact, ceteris paribus, subsidies are beneficial in the "virtuous" models provided they do not reach congestion level, after which public intervention must tend more towards expanding this limit rather than expanding use, otherwise there is the risk that the rate of development of the city will contract. A subsidy for artistic activity alone is not only insufficient but it may well be counter-productive. Investment is necessary for a decongestion of the city. This implies different requirements for the different models.

In model *A1*, congestion is a function of the growth of the income of the city. The problems are general ones and require general solutions.

The subsidies have a relative influence on the increase of congestion. Intervening in the artistic sphere alone risks being useless in that it is only marginal in reference to the general problem.

The case of *A2* is different. Here, congestion is the function of aggregate demand. Subsidies therefore directly accentuate the growth of external economies generated here directly by the aggregate profits re-invested in culture and may forestall or in part counter that situation. It is possible for "decongestioning" to be a conscious part of the investment in culture in the private sphere as well. This investment, and this should not be overlooked, is "instrumental" in the growth of the city. In the case of *B1* as well, the private sector's specialised investment in culture tends, by definition, to de-congest the use of the city, as its aim is to maximise net profit, as in the preceding case, and, moreover, it has an increasing productivity. Congestion here is a direct function of artistic endowment. Subsidies can increase it directly but it is a greater investment in art, which, ceteris paribus, is proper to this model, that counters it.

In conclusion, in all these cases, subsidies may be un-harmful, yet they are insufficient in resolving cities' problems. They must be coupled with an appropriate conservational approach to the economy of the city. The existing "virtuous" model must continue to function, even in the presence of congestion.

In the case of a "harmful dependence" of the economy of the city on its culture (*B2*), however, subsidies are counter-productive. In fact, they provoke an accentuation of the implosive tendency because they always boost the use of the existing cultural endowment and therefore its destruction. There is only one exception, in which the process of "dis-investment" can be overturned so as to re-produce, and conserve the culture used as "input" by the productive processes of the city. That is, it is possible to lead the city either to produce culture by re-investing its productive process, or to transform it into a museum. In this way, the function of cultural investment is shifted upward towards the left in the quadrant until it is made "virtuous".

In the opposite case, instead of or as an integration to subsidies, rationing is made necessary. It is therefore necessary to reduce, rather than increase, demand and supply. The cities belonging to the "neutral" model also have the problem of maintaining or re-inventing their role, and a subsidising of their culture is perhaps a necessary but insufficient condition for a slowing down of the process of decay.

Finally, the museum-cities can certainly be subsidised, but here too this policy alone presents difficulties. In fact, lowering the curve of demand which derives from this cannot improve the overall position (which is maximal by definition) but risks producing overbooking and related phenomena of congestion. A specialised public investment policy aiming to shift the production limit or the investment curve to the left if it has fallen to the right (Figure 5.3) would be preferable.

6. SOME CONCLUSIONS

In conclusion, it is not possible to speak indiscriminately about the crisis of cities of art, but it is necessary to define each of the cities according to cause and typology, otherwise it is not possible to grasp their problems in the deep. Nor can generalised policies of subsidies for these cities be defined without running the risk of producing damage which is even greater than the problems being faced.

There are particular situations where the historical solution to the problems of congestion is partly automatic. This is because of the inter-temporal tendency of net aggregate profits and the external economies they generate in terms of cultural and artistic endowment to increase along with the congestion itself (cases *A2* and *B1*). There are, however, situations in which only exogenous interventions allow for an overcoming of the crisis, whose characteristics are more general than the overall horizon being looked into here (cases *A1* and *C1*). These interventions require both resolving the general problem of the congestion of growth (case *B2*) and sustaining or giving back to the city its original non cultural role if it has already lost it (case *C1)*.

A pure subsidising of the supply, of the demand or of an artistic investment is not sufficient. Moreover, if applied around the limit of congestion it may provoke involution processes. This is particularly true in the case of the model of harmful dependence of the economy of the city on its culture (*B2*) and in that of the museum city (*C2*) because of the risk of an acceleration of the destruction in the former and of overbooking in the latter.

In all cases, it is necessary to intervene on the aggregate net profit curve and therefore on the congestion limit or on the efficiency of the cultural sector per se. This is possible by specialising city investments in culture (cases *C2* and *B2*) and producing new culture through the use of existing culture (case *B2*), thus turning a negative dependency into a positive one. The only alternative, like it or not, is rationing.

Faced with conclusions such as these, which are not at all the usual conclusions, the delegated Italian decision-makers (bureaucrats and politicians) demonstrate two types of reactions. There are those who are scandalised, maintaining that "art has its own economy which cannot be understood by the economy of art". What they mean is that art belongs to everyone and that it is not "moral" to ration people's access to art. Moreover, this access must be promoted as much as possible through subsidies.

The mere mention of prices or the rationing of the cities of art is for them the violation of an ethical imperative. Ethically, these people may well be right; economically, however, they are not. They are comparable to the overly-compassionate doctor who risks killing his patient. These people will not hear of such things as "additional prescription fees" for cities of art. It is not possible to speak of discriminatory policies regarding prices applied to these cities. It is not possible to tax, with discrimination, the proprietors of culturally important buildings in these cities. It is not possible to make a profit out of a productive use of these as means of financing their conservation etc.

There are those, however, who feel that it is proper to give an economic structure to the problem, but the conclusions to this analysis are considered to be politically impracticable. Thus, in the best of hypotheses they fob the problem off onto superior decision makers who will have to define policies. For the most part, they simply do not make any decisions themselves.

7. AN ILLUMINATING QUOTATION

Suspended between these two reactions, Italian cities of art risk becoming a "Nowhere land", present only in intellectuals' Utopian reveries. What we would really like is that it might be possible for people to say of Venice or Florence in the year 2100 what the narrator of News from Nowhere says of Oxford. "By the by", I said, "may I inquire if Oxford is still a centre of culture?". "Still?" he exclaimed smiling. "It has reverted to its best traditions: you can well imagine, therefore, just how different it is from what it was in the 19th century. Now, what is taught in Oxford is real culture, knowledge cultivated for a love of knowledge, not the commercial culture of the past". (William Morris: *News from Nowhere*, 1981).

NOTES

1. See Mossetto (1990) and (1991), where Venice and its culture are analysed as a complex cultural good belonging to the general typology of the "performing arts".
2. Goldthwaite (1984), 113-117 and 117-123.
3. Goldthwaite (1984), 575-586.
4. *Ibid*, 127-131.
5. An efficacious description of these rites and their cultural implications can be found in Heers (1988), 16-50, 87-123 and 162-171.
6. See Panella (1984), 294-302.

REFERENCES

Goldthwaite R.A. (1984), *La Costruzione della Firenze Rinascimentale*, Il Mulino, Bologna.

Heers J. (1988), *La vita quotidiana nella Roma pontificia ai tempi dei Borgia e dei Medici*, Rizzoli, Milano.

Mossetto G. (1990), "A Cultural Good called Venice", in Kakee A., Towse R. (eds.), *Cultural Economics*, Springer-Verlag, Berlin.

Mossetto G. (1991), *L'analisi della domanda di beni artistici*, Note di lavoro 1/92, Dipartimento di Scienze Economiche, Università di Venezia.

Mossetto G. (1992), "The economics of a city of art: a tale of two cities", *Ricerche economiche*, 2-3.

Panella A. (1984), *Storia di Firenze,* Le Lettere, Firenze.

6. THE PROVISION OF AMENITIES BY AGRICULTURE AND RURAL TOURISM

FRANÇOIS BONNIEUX AND PIERRE RAINELLI

INTRODUCTION

The appeal of rural areas for recreation and tourism is directly linked to the increase in income and the urbanisation process. Higher incomes encourage the demand for environmental quality, since they are accompanied by higher education, increasing the awareness of pollution and its harmful effects. The industrialised world, and the consequent urbanisation, explain the appeal of extra-urban environments in the context of a push-pull model of motivation (Pigram, 1993). For urban dwellers, a rural environment appears to support various opportunities to experience compensatory alternative surroundings and cultural or recreational activities.

On the other hand, the economy of rural France, as in many developed countries, has been disadvantaged by structural changes and economic dislocation since the beginning of the sixties. At that time a double movement occurred with greater regional and on farm specialisation and greater regional concentration leading to land abandonment and desertification.

These problems coupled with recognition of the potential contribution of tourism to rural economic development have encouraged the development of rural tourism. But such a tourism needs a provision of environmental goods which are only produced by sustainable agriculture.

The first section provides a set of tools to examine both supply side, (through land use patterns) and demand side (type of accommodation, number of tourists etc.).

The second section draws attention to the relationships between rural tourism and sustainable agriculture since the greater part of rural amenities derive from agriculture.

The third section considers a general framework to assess rural amenities according to the nature of these amenities and how they can be captured by the local communities.

The fourth section presents an illustrative example on sportfishing and its economic impact on a region, Lower-Normandy where surveys have been conducted on salmon and sea-trout anglers.

1. FRENCH RURAL TOURISM

Before examining tourist frequentation in rural areas we give some general information about land use patterns and rural areas in France.

1.1. Land use patterns and rural areas

Substantial territorial imbalance did not arise in France until the 19th century, with the development of industrialisation and the following urbanisation. The city of Paris and its immediate suburbs, which at the time of the French Revolution in 1789 accounted for only 2 percent of the French population, had seen its share doubled by the midnineties. From then on, considerable growth occurred in the Paris area at the expense of the rest of France. From 1850 to 1950, the population drain was specially severe in upland areas. Furthermore 20 per cent of the French population is now concentrated in the Paris area defined in a broad sense (Ile de France), and nowadays some 80 percent of the population lives in cities.

Land use patterns have gradually settled into a few major categories: urbanised areas, farmland where areas of intensive farming are juxtaposed with marginal rural areas, and forests. Major European share arable land is found in France in 60 percent of the farmland, whereas farmland accounts for 55% of the total territory. The share of agricultural area under low-intensity farming systems, which provide a large part of rural amenities, is 25% (Bignall and Mc Cracken, 1996). Forest, which covers 28 percent of the territory is steadily rising: the area of agricultural or marginal land given over to forestry is increasing by 20,000 to 30,000 hectares a year. Timber production remains a key objective, though forest management, chiefly in public forests, is now taking account of conservation of wildlife and forest habitats, hunting, tourism and recreation.

The influence of agriculture is highly visible in the quality and the diversity of French landscapes and in the concept of nature as garden. However, when in the sixties agriculture became a sector closely linked with input suppliers strengthening the productivity of the soil, a disruption to the generations of farmers previously practising agricultural techniques in harmony with the environment occurred. The associated developments, land reorganisation and the abandonment of traditional mixed farming for large-scale growing of cereals and other industrial crops have had an adverse impact on the environment and the conservation of habitats and wildlife.

Over the same period both the rural population and the area under cultivation have fallen, while urban development and afforestation have increased. But local conditions are highly diverse as demonstrated by mountain areas. France adopted around 1880 a policy to restore plots of mountain and hill land to limit the effects of overfarming. This overfarming, which was due to over population caused extremely harmful erosion in the most fragile areas, having serious consequences on the environment. The population drain to Paris and other large cities which occurred at the end of the nineteenth century and the first half of the twentieth century, was especially severe in upland areas. After the second world war the policy objective became to ensure the continuation of farming, thereby maintaining a minimum population level and conserving the countryside.

While the upland agricultural exodus has not completely stopped, it is now more under control. Since 1970 it has taken place at virtually the same pace as the flight from lowland farms, whereas between 1955 and 1970 the drain on the farm population was 1.5 times faster in mountain areas than elsewhere (see Table 6.1).

Table 6.1. Rate of change in the number of farms (per year)

	Mountain areas	Non-mountain areas
1955 - 1976	-3,6 %	-2,4 %
1970 - 1988	-2,6 %	-2,4 %

Source: General census of agriculture.

Nevertheless a closer analysis at the massif level (The Vosges, Jura, Northern Alps, Southern Alps; Corsica, Northern Massif Central, Southern

F. Bonnieux and P. Rainelli

Massif Central and Pyrénées) shows a contrasting situation for the recent period (see Table 6.2).

For the period 1982-1990 Table 6.2 indicates that the overall population of the massifs began to grow again in the eighties. In fact, the growth was heavily concentrated, especially in the Alps, thanks to the rapid development of tourism, mainly winter tourism based on skiing. Except Northern Alps and Vosges, the massifs have a low density, e.g. Jura and Massif Central around 50 inhabitants per square kilometer, or a very low density e.g. Southern Alps, Corsica and Pyrénées (less than 30). Concerning the worforce Table 6.2 shows that farming never reaches 16 per cent of total jobs, with the major part of employment coming from services.

Table 6.2. Demographic and economic characteristics of the French massifs

	Population in 1990		Change 1982-90	Jobs within the area, 1990			
	Total	Density (per Km^2)	% per year	Farming %	Industry %	Construction %	Other %
Vosges (All)	582,002	79	-0.01	5.0	40.6	8.0	46.4
Jura (All)	500,832	51	0.88	7.9	38.0	7.0	47.1
Northern Alps (All)	1,652,890	85	1.10	3.4	25.3	8.4	62.9
Southern Alps (All)	560,981	27	2.01	8.6	13.0	11.7	66.7
Corsica (All)	250,371	29	0.52	8.3	7.3	11.6	72.9
Massif Central (All)	3,700,158	47	-0.11	11.3	24.8	7.3	56.5
Pyrénées (All)	479,310	27	-0.01	14.0	17.4	8.3	60.3
FRANCE	56,610,938	104	0.52	6.3	23.5	7.7	62.5

Note: "All" includes all areas having broadly to do with mountains, less-favoured areas and areas within the mountain mass that are outside the less-favoured areas.
Source: INSEE, special breakdown of the 1990 population census.

1.2. Tourist frequentation in rural areas

As in most European countries, France is experiencing a decline in mass tourism which benefits rural tourism. The concentration of people in both time and space has cumulative effects on scare resources and services such as land, freshwater and sewage treatment. The overcrowding occurring at peak periods leads to traffic congestion, noise, air and water pollution. All these disamenities have a negative impact on the welfare of tourists who suffer externalities produced by themselves. This internalisation of the disamenities explains a growing demand for another type of tourism based on countryside resources which is not completely environmentally friendly, but which is not perceived by tourists and local people as depreciating their enjoyment and appreciation of the area.

The fact that the countryside is more and more popular is observable through the increase in the number of people involved in hiking. Occasional hikers numbered 1.9 million in 1984 and 3.9 five years later, whereas the number of regular hikers grew from 0.840 million to 2.200 million during the same period (IFEN, 1994). Nowadays the number of national park visitors is over 2 million a year. A recent survey found that sixty percent of the French want recreational activities in quiet places.

Statistics on overnight stays help to measure the growing importance of rural tourism. The French statistics on overnight stays include all travel by people to destinations outside the place they normally live, for any accommodation establishment, free or not. Travels for professional, educational or health purposes is excluded. With regard to total overnight stays the 1994–1995 survey on the vacations of the French indicate that a countryside destination is the most attractive. When French people over 15 years old are asked about the places they have been to, 39% respond they have had at least one vacation in the countryside. However the number of countryside overnight stays is lower than the average: 4.7 days versus 5.6. From 1991 to 1995 the number of countryside overnight stays grew twice more than the total number of stays: 11.5 percent versus 6 percent (INRA-INSEE, 1998).

Compared to "classic" tourism, countryside tourism is specific since free accommodation represents 80 per cent of the total stays versus 60 per cent for overall holidays. For instance, the share of family reception accounts for 46 per cent of countryside holidays in 1994–1995 versus 34 per cent of overall stays. The other important free accommodation comes from second home use which represents a quarter of countryside overnight stays. The main paid for accommodation is camping with 8 percent. Farm accommodation and bed and breakfast represent only 4 percent (INRA-INSEE, 1998).

If we turn now to the supply side, we can define the characteristics of the countryside capacity of reception. In 1988, as indicated in Table 6.3, the total reception capacity for tourism reached 19 million beds, of which 11 million were located in rural areas (58% of the total). In rural areas, as well as in the whole territory, weekend homes were predominant (over fifty percent), followed by camping (about 20%). The most specific countryside accommodation, reception in farms ("les gites ruraux") and bed and breakfast accounted for a small percentage: 1.5% and 0.3% respectively. Nevertheless these two types of accommodation have increased substantially: between 1980 and 1988 the supply was multiplied by 4. From 1991 to 1995 the increase reached 46%.

Using the last general census of the population it is possible to have more information on second homes (see Table 6.4). This type of dwelling is very popular in France since it doubled from 1968 to 1990, representing 12.5% of total dwellings in 1990. But compared to the results in Table 6.3 we have a more substantial number of second homes in the rural area (56.8% versus 52.0%). This discrepancy is due to the fact that the 1988 results were issued from a municipal inventory for which the basic units were not quite similar.

Table 6.3. Tourist reception capacity according to type of accommodation and type of territory in 1988

	Hotels	Bed and breakfast	Furnished rooms	Reception in farms	Camping	Weekend homes	Other	Total % N° (000)
Rural areas	4.3	0.3	13.1	1.5	21.0	52.0	7.9	100 (11059)
Total France	7.8	0.3	10.8	1.5	19.0	54.4	6.7	100 (19048)

Source: INRA-INSEE, 1998. Contours et caractères. Portrait social: les campagnes et leurs villes.

Table 6.4 indicates that the increase in second homes is more significant in rural zones than in urban ones. We can also notice that among the various rural areas, remote areas experienced the most important variation. These latter represented in 1990 about 30% of the total French weekend homes.

Concerning farm reception, the 1979 Agricultural Census shows that 12,300 farms had tourist activity: riding schools, organisation of private hunts, game breeding, self-catering accommodation, bed and breakfast, nature discovery. In 1995 this number grew to 16,500 with new farmers better educated and more professional. On some farms agritourism is coupled with farm house produce.

Table 6.4. Evolution of the number of weekend homes according to the type of area

| | 1990 | | Evolution from 1968 to 1990 | | |
	N° (000)	%	68-75	75-82	82-90
Rural areas	1,600	56.8	+ 45	+ 40	+ 23
of which					
weakly influenced by	*473*	*16.8*	*+ 41*	*+ 28*	*+ 16*
metropolitan areas					
Rural poles and	285	10.1	+ 52	+ 46	+ 29
periphery					
Remote areas	842	29.9	+ 46	+ 48	+ 25
The whole France	2,819	100	+ 37	+ 35	+ 24

Source: INRA-INSEE, 1998. Contours et caractères. Portrait social: les campagnes et leurs villes.

2. RURAL TOURISM AND SUSTAINABLE AGRICULTURE

For many rural communities new employment opportunities in recreation and tourism are off-setting the loss of employment in agriculture and other natural resource industries. This rural development is based on the environment considered as a rural resource. Because farming occupies so much of the rural territory, its role in managing rural space and shaping the rural landscape is fundamental. The role of agriculture in providing rural amenities which can be used to increase development opportunities in rural areas can be seen through Figure 6.1.

In the upper part of Figure 6.1 marginal private cost of agriculture and marginal social cost are expressed according to the level of intensification. This latter is measured by a simple indicator Q which is the output per hectare. The marginal social cost is U shaped whereas marginal private cost is a simple increasing linear function. Consequently there are two intersections of the curves: A and B corresponding to the intensification levels Q_1 and Q_2. The difference between the two curves, the shaded areas, represents the balance between negative and positive externalities produced by agriculture. The lower part of the figure presents this balance.

The balance between negative and positive externalities is negative in the left part (OQ_1) and on the right part (OQ_2). Between Q_1 and Q_2 agriculture generates a net flow of amenities and between these levels of intensification it can be defined as sustainable.

There are negative externalities for OQ_1 and beyond Q_2 for opposite reasons. The negative externalities corresponding to OQ_1 are due to the withdrawal of farming in less favoured areas such as mountains. In the Alps, for instance, the abandonment of pastures and grassland areas with difficult orographic conditions has negative consequences. When permanent forage crops situated in sloping zones are not mown or cut regularly there is an important risk of erosion. Moreover, in winter the snow mantle is not stabilised and avalanches can occur. In the Mediterranean area, when pastures are not grazed as they revert to scrub the risk of fires also increases.

The withdrawal of farming also affects biological diversity and landscape aesthetics. After a few years of scrub invasion, its growth can block out vistas, narrowing the horizon and increasing monotony. In the Vosges, a mountainous massif hit especially hard by agricultural exodus, the collapse of traditional farming has led to piecemeal spontaneous afforestation, primarily by spruce trees, which first affects former communal grazing, tilled plots and sometimes even meadows. Rough grass develops in pasture that is grazed occasionally, giving high but uneven growth, and bushes start to appear. The natural development of trees and scrub eventually produces rough woodland. The clear, orderly design that had once shaped the countryside is lost, and the landscape is closed off. This tends to isolate the remaining inhabitants, eroding the quality of the setting in which people live (Gagey and Rainelli, 1996).

The negative externalities beyond Q_2 come from the adoption of industrial processes by modern agriculture, with mechanisation and "high input farming". Agriculture in today's economy is often completely delocalized with regard to its commodity end-use, and it is no longer dependent on local resources for its operational inputs. The ways in which intensification has damaged the countryside may be analysed under a number of headings.

Regarding cropping patterns, the most significant aspect for the environment is the shift between arable land and permanent grassland. From 1970 to 1995 the former increased by 1.2 million hectares, whereas the latter decreased by 1.5 million hectares. If we consider the principal categories of crops there is a dramatic increase in wheat from 3.7 million hectares to 4.3 and to a lesser extent in maize (1.4 to 1.7 million hectares). The development of monoculture has created a monotonous landscape which most people do not find attractive, and high land-use intensity with its consequences: destruction of biotopes and loss of biodiversity, and the contamination of surface and ground waters.

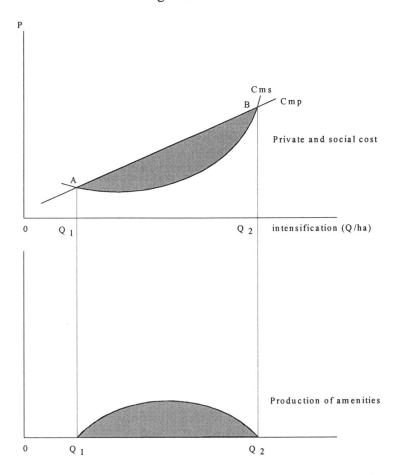

Figure 6.1. Agricultural sustainability and provision of amenities

Comparable developments have taken place in the livestock sector characterised by a dramatic increase in intensive animal husbandry, notably pigs and poultry. Local effects of this intensive farming concern visual amenity since high-quality rural landscapes are damaged by the erection of grain silos and modern agricultural buildings for animals. Moreover, the quantity of manure produced by pigs and poultry, combined with the other sources of nutrients, cause large quantities of nitrates and phosphorous to run off into rivers or aquifers. This oversupply causes eutrophication of slow-running and surface waters. Eutrophication also occurs in estuaries and marine waters leading to the disruption of ecosystems, with algae growth disturbing the oxygen balance, which subsequently results in the death of fish. In addition, algae emit bad odours when they become distorted.

Otherwise, intensification and specialisation have come about through major structural changes, i.e. a rapid increase in the average farm size (from 13 ha in 1955 to 35.5 ha in 1995) involving the substitution of machines for manual labour. The number and power of machines used per square kilometer of arable land has continued to rise, leading to soil compaction and erosion. Furthermore, the ever increasing size of machines calls for wider plots which are obtained through land amalgamation and plot consolidation. Since 1945 15 million hectares have been consolidated, half of the total farmland. The consolidation of land holdings has been very important in Northern France, mainly in Bassin Parisien where cereal farming dominates, but also in the East (Alsace) and in the West (Bretagne and especially Morbihan). Puy de Dôme in the Massif Central has also seen significant land consolidation.

The most important effect of plot consolidation on the landscape and nature is the elimination of features such as hedgerows, groves and isolated trees. The best index of this loss of quality is hedgerow removal, known thanks to the national survey of forestry (Pointereau and Bazile, 1995). In 1963 five "départements", all located in the western part of the country, contained 25 per cent of the total hedgerow length (Côtes d'Armor 102,000 km, Finistère 79,120 km, Vendée 51,100 km, Manche 46,736 km, Loire-Atlantique 41,160 km). In these five "départements" the total removal of hedgerows between 1963 and 1980 reached 200,000 km.

The destruction of hedgerows in the western part of France leads to the disappearance of particular enclosed landscapes named "bocages" the result of centuries-old human use. These cultural landscapes, the term "cultural" characterising the distinctive interrelationship between nature and people (Stanners and Bourdeau, 1995), are very much appreciated by tourists and countryside lovers because of the richness and diversity of the habitats and scenery. The "bocages" are associated with traditional agriculture based on pastures and low-input farming.

The changes in the landscape caused by the development of monoculture and the dramatic increase in intensive rearing reduces the supply of rural amenities. This reduction has negative consequences on tourist welfare, as demonstrated by a research conducted in Brittany using the hedonic price method (Le Goffe, 1997). The objective was to find the policies for rural tourism by identifying the negative and positive externalities which influence the renting price of a sample of 600 rural cottages located in Brittany.

Each cottage is described using three main categories of attributes: intrinsic (capacity, comfort), geographic (how far the cottage is from Paris and the sea) and environmental (share of woodland, permanent grassland, cereals, forage in the territory of the "parish" and density of pigs and poultry in the "parish"). Concerning the environmental variables, the renting price is positively influenced by the share of permanent grassland, an indicator of extensification, and negatively influenced by the share of forage crops and density of pigs and poultry. When the share of forage crops increases by one point the renting price decreases by FF 5. When the density of pigs and poultry, expressed in livestock units, increases by one livestock unit, the renting price decreases by FF 120. The average price is FF 2,000.

If we now consider the medium level of intensification corresponding to the interval $Q_1 Q_2$, the nature of environmental goods with regard to tourism will differ according to region and type of farming. The relationship between local conditions and land-use on the one hand, and tourist resource on the other can be shown graphically (see Figure 6.2 derived from Figure 6.1).

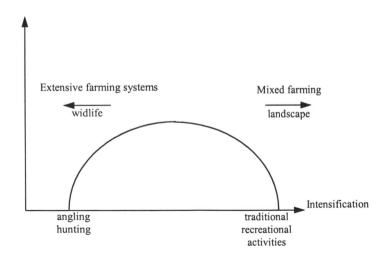

Figure 6.2. The relationships between type of farming and tourist opportunities

The precise shape of the curve of Figure 6.2 will vary according to regions and types of farming. This relationship implies the existence of important joint-product relationships between agriculture and provision of environmental goods which are inputs for tourist activity. However, the major question is the degree to which the rural amenities linked to sustainable agriculture can be used to produce marketable services. Where environmental values are high the first way in which markets permits the commoditisation of rural amenities is through the sale of land and buildings for second homes. Other forms of marketisable environmental product exist which are examined in the next section.

3. GENERAL FRAMEWORK

The close relationship between sustainable agriculture and rural amenities is used as a development tool for the areas in which they lie. Rural areas offer the whole of society, mainly urban centres with high income, a valuable service which are partially paid for through countryside tourism and other forms of commoditisation.

From an economic point of view the different amenities represent a value to some or to all individuals, just as other commodities valued by the market. The different utilities derived from amenities allow us to differentiate them according to their type, their local impact and their possible commoditisation for agriculture. An attempt is made in Table 6.5 on the basis of mountainous zones. Amenities are hierarchised according to the various values: user value and ecological value.

Concerning user value, the utility or the value people derive from the underlying amenity is linked to the direct physical use of the corresponding amenity, or indirectly linked. When there is a direct use value through the existence of a "terroir", a circumscribed territory having unique soil associated with traditional agricultural and processing practices, commoditisation by the market is simple. The existence of a "terroir" allows the use of "Appellation d'Origine" labels chiefly for wine and cheese, but nowadays also for other products such as vegetables or olive oil. In the case of cheese, there is a famous one in the Alps, the Beaufort, corresponding to a specific area with specific obligations to feed the cows and process the cheese.

Table 6.5. Possibilities of commoditisation for tourism purposes of services rendered by mountain and hill farming

	Source of amenity	Local impact	Value to agriculture	
			Market role	Form
Direct use	Existence of a "terroir"	Products entitled to a special label "Appellation d'Origine"	Yes, directly	Premium prices for products
value	Quality of landscapes	Space management special farming practices	Yes, partially via agri-tourism, bed and breakfast	Provision of services landscaping contracts
Indirect use value	Protection against flooding and erosion	Land maintenance and favourable farm practices maintenance of terraces	No	National and European system of aid
	Biological diversity)	Presence of particular ecosystems	No	Maintenance subsidies
Ecological value	(fauna and flora habitat richness	hunting and angling	Yes, partially via tourism	Special contracts with farmers and landowners
	Contribution to balanced land-use planning and development	Presence of a certain agricultural population	No	Grants to retain farmers

Note: "terroir" can be defined as a circumscribed agricultural area having unique soil associated with traditional agricultural and processing techniques.

The higher the intrinsic quality of the "terroir", the more pronounced the price effect. In other words, "differentiation of agricultural products according to quality and origin, seems to have a series of positive consequences not only for farmers' income, but also for the environment and rural amenities" (Merlo, 1996). The quality of landscapes also provides a collection of services which can be partially commoditised via agri-tourism.

But use value can be indirect since sustainable agriculture in mountain areas with pastures and terraces avoids landslides, erosion with offsite damage, and flooding. The utility people derive from favourable farm practices, or more generally from the presence of a certain farm population, is linked to the amenities produced, but in an indirect way. For such positive externalities there is no market. From a social point of view, the Paretian optimum needs that farmers should be compensated for the provision of these services by an amount which is equal to the sum of consumers' marginal utility. Theoretically this amount must be paid by the consumers who benefit from these non-market services. In fact, compensation to offset natural handicaps is given by governments using the structural measures (Directive 268/75) foreseen by the EU. In consequence it is the taxpayer who pays and not the beneficiary of the services provided.

Sustainable agriculture in mountainous zones is also a source of ecological value through the prevention of depopulation in less favoured areas and the development of balanced and viable rural areas. For such value there is obviously no market. The conservation of biodiversity provides two types of amenities. The first places the emphasis on variability or heterogeneity within species, between species and ecosystems. This variability appreciated by the species richness can result in particular ecosystems which present scientific interest. Genetic resources can have an option value, an existence value. On the other hand, the richness of the habitat can give marketable services when hunting, angling or bird-watching are possible. Mushroom picking or berry picking can be a source of revenue for the local population.

Table 6.6 shows that the direct or indirect remuneration of an amenity does not capture the total value of the amenity. This is particularly clear with regard to landscapes. Several studies have been conducted to estimate the value of agricultural landscapes by means of the contingent valuation method. Even if the policy and spatial attributes involved differ it is possible to compare the results. In the case of Sweden, willingness to pay to maintain an open landscape for the whole country is considered (Drake, 1992), whereas other studies address the landscape issue at the local level (Le Goffe and Gerber, 1994; Bonnieux and Le Goffe, 1997). The British studies are related to Environmentally Sensitive Areas, or National Parks, and they consider the regional level.

Table 6.6 indicates that for a given geographical level the annual willingness to pay is of the same order of magnitude. The large amount obtained by Bateman et al., 1995 is due to the specificity of the Norfolk Broads which are under threat from flooding. In consequence the change is presented to people surveyed as irreversible, and willingness to pay is high.

Production and conservation of beautiful landscapes have a value, as expressed in Table 6.6 using the contingent valuation method, but they do not have a price on a defined market. The proportion captured by rural tourism is very small. This is due to the fact that landscape is a public good. It is impossible to exclude some people from consuming it. Jointures of supply is the other characteristic, meaning that the marginal cost of letting additional people consume the commodity is zero.

Table 6.6. Willingness to pay for various agricultural European landscapes (in 1994 ECU)

Author	Country	Geographical level	Willingness to pay
Drake, 1992	Sweden	national	99 per person/year (general public)
Pruckner, 1995	Austria	national	0.30-0.78 per person/day (tourist)
Garrod et al., 1994	UK	regional	22.4 per hsd/year (resident) 15.1 per hsd/year (visitor)
Garrod and Willis, 1995	UK	regional	35.2 per hsd/year (resident) 24.9 per hsd/year (visitor)
Willis and Garrod, 1993	UK	regional	33.7 per hsd/year (resident and visitor)
Bateman et al., 1995	UK	regional	100 - 187 per hsd/year (resident)
Le Goffe and Gerber, 1994	France	local	28.4 per hsd/year (resident)
Bonnieux and Le Goffe, 1997	France	local	30.9 per hsd/year (resident)

On the other hand, an amenity such as hunting can be viewed as a private good since there is excludability and rivalry in the use of this resource. Between these extremes live the majority of goods, where exclusion would be possible and where congestion causes a diminution in value at certain levels of use. Following Whitby (1990) we can classify some rural amenities in terms of their rivalries and excludability (Figure 6.2).

Table 6.7. A classification of some rural amenities

	rival	non rival
excludable	shooting, fishing	public monuments (castles)
non-excludable	foot paths and bridleways	landscapes, lakes

For some of the values and benefits linked to sustainable agriculture, the beneficiaries are local people. But in many cases the geographic spread of amenity benefits is important and a large part of rural amenities accrues to individuals living outside the area where conservation costs are incurred, leading to "territorial spillovers". When the services are free, or not properly priced, there is a territorial externality. According to economic theory public goods and externalities are market failures leading to an inefficient allocation of goods. To restore the Paretian optimality we have to internalise these unpriced public amenities. Internalisation gives an incentive to the provision of amenities by preserving existing amenities, or promoting new ones.

With territorial spillover it appears that the issue of property rights is crucial with regard to public amenities. Society accords implicit property rights to rural people and especially to farmers whose role as guardians of the countryside is firmly acknowledged. Direct payments foreseen by regulations 2328/91 and 2078/92 are justified because farmers accept constraints on their activity resulting in a reduced level of negative externality. These schemes can be interpreted as a reallocation of property rights, the payment being the price for purchasing or renting rights by public authorities.

To avoid a sub-optimal level of provision of public amenities a certain number of policies are possible. According to Hodge (1993) four categories of responses exist: stimulating the supply of amenity, raising funds to fund environmental protection, regulating the levels of amenity consumption and stimulating markets for the provision of amenity.

Stimulating supply

Policies should tend to offer positive incentives to farmers providing public amenities by encouraging the generation of amenity benefits or discouraging the destruction of amenities. Theoretically, the best way to stimulate supply is to place a price on the valued attribute of amenity. This price is not so easy to determine as shown by Table 6.5 on the values of agricultural landscapes. Pricing can be linked to some proxy for the level aspect of amenity. For instance, the enhancement of landscape amenity in "bocage" region can be based on the restoration of hedgerows and hence the length of existing hedges.

In practice, agri-environmental measures associated to the traditional measures of agricultural support, are aimed to "compensate farmers for any income losses caused by reductions in output and/or increases in costs and for the part they play in improving the environment".

Raising funds to promote the provision of environmental goods

Funds to protect or enhance positive externalities can be obtained directly by soliciting people or economic agents benefiting from the quality of the rural environment, either consumers or firms. This is more common in Northern European countries than in Mediterranean countries. Nevertheless examples exist in France in the field of marine ecosystems with the Cousteau foundation, or the Paul Ricard foundation where funds are given by a firm to promote environmental research.

Funds can be obtained by charging users, such as visitors to Natural Parks in the United States who have to pay an entrance fee. Such a policy seems very difficult to implement in France. Surveys have been carried out in a peri-urban forest, near Rennes in Brittany, to measure the amenities provided to urban people visiting it. The willingness to pay amounts to FF 110 per household per year, but there is a unanimous refusal of tolls or entry fees. However there is acceptance of raising funds by increasing local taxation (Bonnieux and Rainelli, 1996). Instead of entrance fees it is possible to raise a charge for car parking.

Regulating uses

Non-price constraints can be used to limit the pressure on fragile sites. They can be based on regulation or making access to a site more difficult by, for example, removing car parking facilities. Commoditisation in order to find the financial resources requires a legal framework which sometimes leads to the modification of property rights, such as the restriction of mushroom and berry picking. In another way the legal framework is useful to protect the "appellation d'origine" products, as the European level Regulation 2081/92 sets out the Protection of Products Origin and Quality. On the same line the Eco-Labels policy can also be considered as promoting rural amenities.

Stimulating markets

Even if the scope for the direct consumption of an amenity to be captured within markets appears to be limited, there are some possibilities. Besides the "appellation d'origine products" the most extensive efforts at commoditisation of rural space relate to the provision of leisure facilities in the countryside. The rapid growth of golf courses is the best example. But golf developments can cause environmental damage where they require removal of earth, disrupt existing hydrological patterns and entail the draining of wetlands to create greens. In this circumstance the development of new markets based on rural amenities competes with the provision of amenities.

4. GAME ANGLING AND REGIONAL DEVELOPMENT: THE CASE OF LOWER NORMANDY

Suppliers of rural amenities are not adequately rewarded when there is a high proportion of non-use values attached to amenities, and when it is difficult to exclude beneficiaries and thus to charge them. If we consider outdoor recreation use values represent a large part of the amenity, but according to the type of recreation barriers to the capture of the value can exist. For example, mountain communities cannot charge for access to mountain roads in the case of mountain walking.

On the contrary, sportfishing represents an interesting opportunity for local communities since among the amenities provided by hydrosystems use value dominates, and because actual exclusion of potential users is possible through licences. The economic impact of these sportfishermen is far from being insignificant, particulary salmon and sea-trout anglers who have a high income level. Our case study is focused on these two categories of anglers in Lower-Normandy. The first subsection presents the general characteristics of the anglers, the second subsection provides information on the expenditure of fishermen and the third subsection estimates the indirect effects of this activity on regional development.

4.1. General characteristics of the anglers

In France freshwater fishing is a widespread leisure activity. A national survey carried out in 1991 by the National Council of Fishing, the Ministry of Tourism and the Ministry of the Environment showed that 5 to 6 million people practise this activity. About 3 million fishermen go fishing at least five days a year. It can be said that recreational fishermen in the broad sense represent about 9% of the population. The annual budget adds up to FF 6 billion corresponding to an annual budget per fisherman amounting to FF 1,300 (Jantzen, 1998).

For the great majority fishing is a proximity activity practised in the area of one's home district (47 per cent). From this point of view there is a great difference to sportfishermen who come from other regions or other countries. There is another big difference involving the people engaged in this activity: about 3,000 come for salmon and 3,300 for sea-trout. A large number of these anglers practise in the western part of France because of the quality of the rivers of Brittany and Lower-Normandy and the vicinity of Paris, at least as regards Lower-Normandy.

Surveys were conducted on rivers using a detailed questionnaire reviewing anglers' characteristics, fishing experience and effort, and also expenditure. The only question for which non-response is significant involves income: 25 per cent refused to answer. Nevertheless it appears that this population is made up of middle or senior executives and professionals. Note that family income is much higher for people living outside the region, particularly for fishermen coming from Paris.

There are some differences between salmon anglers and sea-trout anglers: the former are older (5 years) and have been fishing over a longer period of time (15 years versus 7 years). The length of the fishing day and the number of trips during the fishing period are greater for salmon than for sea-trout anglers. This corresponds to different behaviour: salmon anglers do not hesitate to spend several days at a time without catching anything, whereas sea-trout anglers are not so eager and visit the river for a half-day only. Round-trip distances are similar for both types of angler, and as a consequence Table 6.8, which gives the distribution of the distance, does not distinguish between the categories.

Table 6.8. Distribution of round-trip distances

< 10 Km	10 - 50 Km	50 - 100 Km	> 100 Km	%	Km
25.2	37.4	12.0	25.4	100	169

Table 6.8 shows that a quarter of sport fishermen come from remote areas, whilst the same proportion are residents. More than 80 per cent use their own car and 15 per cent use a friend's or their family's car.

4.2. Expenditure of anglers and willingness to pay

Basic data were obtained thanks to on-site surveys run during the 1990 fishing season with a sample of 350 anglers. Data have been updated using the results of another on-site survey conducted in Brittany in 1995 with a sample of 176 salmon anglers (Porcher and Brulard, 1998). The structural characteristics of the two samples of salmon anglers are very close, so it is possible to match the results.

The cost of a fishing season including transportation costs, food, lodging, fishing and depreciation of equipment amounts to FF 10,670 for salmon anglers in 1995. For sea-trout anglers the cost of a fishing season is less elevated, about 50 percent less (FF 6,200). The cost of equipment, including reels, rods and lines totals FF 6,950 for salmon anglers and FF 6,425 for sea-trout anglers.

Because of the economic impact of lodging, catering and transport on the local communities, Table 6.9 details this information for the whole sample. The total expenses for lodging, food and transportation reach FF 6,510 per fisherman and per fishing season. Transportation costs represent 60 per cent of the total. For resident people expenses amount to FF 3,726 versus FF 10,000 for outside fisherman.

Table 6.9. Detail of lodging, food and transportation costs (FF per fisherman)

Lodging		Food		Transportation costs	
Hotel	275	Restaurant	751	round-trip	3,707
Camping	88	Other	573	on-site costs	534
Other	222				

At the beginning of the nineties there was a management scheme in place whereby salmon catches were restricted to four salmon per angler before June 1st plus two after. This scheme was considered to be inconsistent since the fishing season started in mid-March and permitted the fishing of spring-salmon which are rare, while drastically limiting the catches of grilse which are relatively abundant. Salmon anglers were asked on their opinion concerning the suppression of the quota system after June 1st. For those agreeing a payment card was used to elicit willingness to pay. The average amount equalled FF 103.

As regards sea-trout angling, the problem was specific to a river for which the building of a fish ladder could give new opportunities to increase the length of banks available for angling. But these banks were privately owned. Thus the contingent valuation study asked people to voluntarily contribute to a fund aiming at 5 km of river banks, knowing they would be entitled to fish free-of-charge for three years. The average amount equalled FF 567 per person yearly.

Econometric models have been built to analyse sea-trout angling demand. First probit and logit models are used to explain the behaviour of fishermen faced with a dichotomous choice, as they are asked to accept or reject voluntary participation in a fund. The dependent variable is a yes/no answer. These are six independent variable:

- two variables (income and years of training) describe the anglers;
- cost of equipment is a proxy for fishing effort;
- number of sea-trout caught the previous season is an indicator of fishing experience;
- a dummy variable takes into account substitute rivers;
- trip distance is considered as an indicator of interest in game angling.

Table 6.10. Sea-trout angling demand, tobit model (asymptotic t values in parentheses)

Variable	Tobit	(t-values)
Intercept	5.810	(17.1)
Site substitute (dummy)	0.270	(1.04)
Trip distance (km)	0.002	(2.5)
Cost of equipment (FF)	0.175	(2.2.)
Catches	0.030	(2.2)
Monthly income (17 classes)	0.033	(1.3)
Years of training	-0.025	(-1.9)
Log likelihood	-34.69	

Source: Bonnieux et al., 1992.

Both models perform poorly statistically, so a tobit model is specified with willingness to pay as a dependent variable and with similar independent variables. The tobit model, which is well adapted to the censored nature of the data, is better because all signs are consistent and t-ratios are greater (see Table 6.10). The demand for game angling rises with income. Anglers who own many reels and rods, and therefore have a high value of equipment, make the greatest fishing effort. The value of equipment positively affects demand. As expected a positive correlation between catch and angling demand is obtained. Trip distance has a positive sign. The negative sign of training years is explained by the fact that this variable is positively correlated with the angler's age. As sport fishing demands great physical effort, senior anglers sometime move to other types of fishing, leading to a demand decrease.

4.3. The indirect effects of sportfishing on the economy of Lower-Normandy

The value of sport fishing is high according to willingness to pay to improve the number of catches, and simultaneously total expenses incurred by anglers are important. In consequence this outdoor recreation makes a significant contribution to the local economy. In this subsection we try to present the aggregate economic effets of sportfishing in Lower-Normandy.

We have first to determine total expenditures corresponding to the direct effects of sportfishing in the region. The data extrapolated from the various surveys available (1990 and 1995) are reported in Table 6.11.

Total direct expenses incurred by sportfishermen who are not resident in Lower-Normandy account for FF 12 million, representing 55 percent of total expenditures. The most important category of expense is transportation (44 percent of the total), followed by logding and food (23 percent). But a large part of the transportation costs of sportfishermen living outside Lower-Normandy are not regional expenditures (52 percent). These outside anglers also buy their equipment in the region they come from, so only 50 percent of this heading corresponds to regional expenses.

Table 6.11. Sportfishermen expenditures in Lower-Normandy in 1995 French Francs

	Lodging and food	Transportation costs	Licence and fees	Depreciation of equipment	Total
Total expenditure	4,933	9,649	3,668	3,606	21,856
of which					
people	3,379	4,759	2,086	1,809	12,033
living outside					

On the other hand, there are other regional expenditures aimed at improving the supply of salmon and sea-trout by building fish ladders and restoring the spawning capacity of the rivers. These investments and expenses are funded by the Region Lower-Normandy and the State within the context of the State-Region Plan Contract. There is also a general scheme to favour anadromous fish. The yearly regional expenses amount to about FF 5 million.

To estimate the indirect regional economic effects the two-digit national input-output table (90 industries) has been used (Boude, 1991). The correspondance between industries and output related to fish tourism is:

02 fishing industry	⇨	cultivated fish
55 construction	⇨	fishing ladders and other similar equipment
57-64 wholesale and retail	⇨	food, gas and fishing equipment
65 car repairs and car sales	⇨	half of transportation costs

67 hotels-cafés-restaurants ⇨ stay expenses minus food

96 non-market services ⇨ expenses related to the National Council of Fishing

Now, fishing expenses including efforts to improve the supply, can be re-arranged according to the related industries. Using adapted economic multipliers and technical coefficients we can estimate total indirect effects (see Table 6.12).

Table 6.12. Indirect economic effects induced by sportfishing in Lower-Normandy (FF 1995)

Industry	Value added (FF 1,000)	Ubtermediate consumptions (FF 1,000)	Total (FF 1,000)
03 fishing	22	5	27
55 construction	2,035	540	2,575
57-64 wholesale and retail	1,410	125	1,535
65 car repairs and car sales	1,250	150	1,400
67 hotels-cafes-restaurants	2,050	390	2,440
96 non-market services	3,100	1,100	4,200
Total	9,867	2,310	12,177

As indicated by Table 6.12 the major part of indirect economic effects induced by sportfishing are concentrated on the industry of non-market services (34 percent), followed by construction (21 per cent), and hotels-cafés-restaurants (20 percent). If we consider direct expenses incurred by fishermen (FF 21.8 million) and indirect effects (FF 12.2 million), the total amounts to FF 34 million, which is significant for the local economy.

CONCLUSIONS

In many rural places structural change in traditional sectors has made economic opportunities scarce, and consequently per capita incomes are well below national average. However, the countryside has much to offer the rest of society in terms of amenities in a general context of growing demand from urban dwellers for serene and quiet landscapes, synonymous with a notion of the "good life", even if it is an idealised vision. In this way, amenities can become assets for rural development and tourism can be a useful means of raising the level of economic activity in regions not well endowed with other resource potential.

The provision of rural amenities is directly linked to the degree of agricultural intensification. On the one hand economic decline and depopulation can harm nature. In particular, cultural landscapes greatly appreciated by day visitors and vacationers decay when agriculture disappears. On the other hand, structural and technological development in agriculture leads to the amalgamation of fields into larger units causing hedgerows, dikes, ditches and other field boundaries to be removed, thus creating monotonous landscapes. In parallel there is an over use of mineral fertilizers resulting in eutrophication of watercourses, lakes and reservoirs. The provision of rural amenities supposes sustainable agriculture since this activity concerns 80 percent of the territory.

Many rural amenities are public interest goods which may be enjoyed or used by many people simultaneously without making them less available to others. Non-rivalry and non-excludability pose a major challenge in terms of linking amenities and development. But when these territorial amenities have a high use-value component local communities can capture a large part of the amenity value. This is very clear for the "terroirs" where existing regulations assure the quality and origin of agricultural products which are labelled. In another way the existence of property rights for game angling, through licences and strict controls, gives rise to an important form of rural tourism. The example of Lower-Normandy shows that the direct and indirect impact of anglers' expenditures, coupled with multiplier effects have the potential to offset economic deprivation. This case also clearly demonstrates the need for rivers of good quality (clear water, sufficient oxygen) suitable for sportfishing. On the contrary, the potentially negative impact of tourism, arising from conflicts with residents and too great a frequentation of environmentally sensitive sites must be taken into account. There is a need for sustainable tourism in fragile areas.

REFERENCES

Bateman I.J., Langford I.H., Turner R.K., Willis K.G. and Garrod G.D. (1995), "Elicitation and truncation effects in contingent valuation studies", *Ecological Economics*, 12, 161-179.

Bignall E. and Mc Cracken D. (1996), "The ecological resources of European farmland", in

Whitby M.C. (ed.), *The European environment and CAP reform policies and prospects for conservation*, CAB international, Wallingford.

Bonnieux F. and Rainelli P. (1996), "Amenities in wetlands and peri-urbain woodland in France", in OECD, *Amenities for rural development - Policy examples*, pp.75-83.

Bonnieux F., Desaigues B. and Vermersch D. (1992), "France", in Navrud S. (ed.), *Prising the European environment*, Scandinavian University Press: 45-64.

Bonnieux F. and Le Goffe P. (1997), "Valuing the benefits of landscape restoration: a case study of the Cotentin in Lower-Normandy, France", *Journal of Environmental Management*, 50: 321-333.

Boude J.P. (1991), "Les effets économiques induits", in Bonnieux, Boude, Guerrier R., *La pêche sportive du saumon et de la truite de mer en Basse-Normandie*, Document CSP-INRA-ENSAR. 78

Drake L. (1992), "The non-market value of the Swedish agricultural Landscape", *European Review Agricultural Economics*, 19: 351-364.

Gagey D. and Rainelli P. (1996), *Intérêt de l'aide à l'agriculture de montagne. Séminaire de l'OCDE sur les avantages écologiques de l'agriculture durable*, Helsinki, 10-13 sept.

Garrod G.D., Willis K.G. and Saunders C.M. (1994), "The benefits and costs of the Somerser levels and moors ESA", *Journal of Rural Studies*, 10: 131-146.

Garrod G.D. and Willis K.G. (1995). "Valuing the benefits of the South Downs environmental sensitive area", *Journal of Agricultural Economics*, 46: 160-173.

Hodge I., (1993). *The contribution of amenities to rural development. Rural amenity: definition, property rights and policy mechanisms*, OECD C/RUR (93) 14.

IFEN, (1994), *L'environnement en France*, Dunod édition, Paris.

Jantzen J.M. (1998). "A national survey on freshwater fishing", in Hickley and Tompkins, *France. Recreational fisheries: social, economic and management aspects*, FAO-UN Fishing News Books, pp.5-9.

Le Goffe P. and Gerber P. (1994), *Coûts environnementaux et bénéfices de l'implantation*.

Le Goffe P. (1996), *Hedonic pricing of agriculture and forestry externalities. The European Association of Environmental and Resource Economics*, 7th conference Lisbon, June, 27-29.

Merlo M. (1996), "Commoditisation of rural amenities in Italy", in OECD, *Amenities for rural development. Policy examples*, pp. 85-95.

Pigram J.J. (1993), "Planning for tourism in rural areas. Briding the policy implementation gap in tourism research", in Pearce D. and Butter W., *Critiques and challenges*, 156-174.

Pointereau P. and Bazile D. (1995), *Arbres des champs: haies, alignements, près, vergers ou l'art du bocage*, Solagro-WWF.

Porcher J.P. and Brulard J. (1998), "An economic analysis of salmon fishing in the Finistère departement of France. Recretational fisheries: social, economic and management aspects", in Hickley and Tompkins, *FAO-UN Fishing New Books*, pp. 200-203.

Pruckner G.J. (1995), "Agricultural landscape cultivation in Austria: an application od the CVM", *European Review of Agricultural Economics,* 22: 173-196.

Stanners P. and Bourdeau P. (1995), *Europe's environment. The Dobris Assessment,* European Environment Agency, Copenhagen.

Whitby M. (1990), "Multiple land use and the market for countryside goods", *Journal of the Royal Agricultural Society of England,* (151): 32-43.

Willis K.G. and Garrod G.D. (1993), "Valuing landscape: a contingent valuation approach", *Journal of Environmental Management,* 37: 1-22.

PART II

INSTRUMENTS AND POLICIES

7. ECONOMIC INSTRUMENTS FOR SUSTAINABLE TOURISM DEVELOPMENT

ANIL MARKANDYA

1. INTRODUCTION

As one of the fastest growing sectors of the economy, tourism has a key contribution to make to sustainable development. At the same time it has been accused, with some justification, of contributing to the degradation of the natural environment and to the destruction of traditional cultural values in a way that is often irreversible. As disposal income grows throughout the world, the demand to visit sites of special significance will only increase. Since the number of such sites is fixed, the pressure will be felt even more acutely than it is today. Already, there are places in Italy where access has to be rationed for tourists and where there are concerns about the degradation suffered by the historical materials from both "normal" visitors as well as vandalism by some tourists[1].

In addressing these problems we need a multiplicity of instruments. Certainly there is an important role for punitive sanctions against offenders to laws protecting natural sites and cultural monuments. Moreover there is a literature on how such sanctions should be designed and operated[2]. This paper, however, does not deal with such regulations. Instead it looks at the use of market based instruments, such as charges and permits, to achieve a socially optimal solution. Essentially the problem is one of externalities; tourists visiting a site impose negative externalities in terms of damage to the natural and constructed fabric of the place as well as imposing congestion type costs on each other (at least after some critical number has been exceeded). This much is, I believe, well understood and the basic principles of the use of market based instruments in the regulation of such environmental externalities has been studies extensively (see, for example Tietenberg, 1997 for an excellent review). The main problems that remain are to design the use of such instruments in a practical context, taking account of the market structures that exist and setting the values for the instruments on the basis of adequate knowledge about the external costs. This is where the effort has to be focussed.

This paper is a modest attempt to understand better what kind of instruments can be used, how the market structure is a relevant factor in determining the levels at which the controls are set, and the key parameters that determine the levels of the market based instruments. As will become clear, there are many gaps in our knowledge that need to be filled if we are to be successful in controlling tourism in a way that puts this important industry onto a sustainable development path.

The paper is structured as follows. Section 2 discusses the main sources of externality in tourism and the lack of empirical estimates of parameters that measure such external effects. It also outlines some of the important "stylised facts" about the tourism industry; and provides a discussion of the private market equilibria for such an industry, including the relation between such equilibria and the social optimum. Section 3 reports on some simulations for an industry that supplies tourism services under monopolistic and monopolistically competitive conditions. Section 4 discusses some further developments that need to be made to this model to analyse more complex, and perhaps more interesting conditions. Section 5 concludes the paper.

2. EXTERNAL COSTS IN TOURISM AND THE STRUCTURE OF THE TOURIST INDUSTRY

What kind of external costs does tourism generate? These fall into two sets. One is related to the damages done to stone, paint, frescos, or to natural rock, coral etc., from the sheer weight of numbers or from vandalism; and the other is the congestion caused by the presence of tourists. It is difficult to appreciate nature or to contemplate a beautiful work of art if the ambient environment is overwhelmed by hordes of tourists doing exactly the same thing.

The external costs of tourism have been discussed and a few estimates exist of impacts of tourism to coral and other materials, but there are virtually no estimates of the monetary values associated with these physical impacts. For analytical purposes, what is needed is not just the average damage per visitor, but the marginal damages per visitor and how these marginal damages change with the number of visitors. In Section 3 some plausible values are provided, but this is clearly an area where further empirical work is needed.

The congestion caused has, likewise, not been assessed in any serious empirical way. The demand functions for tourism have been estimated (e.g. Crouch and Shaw, 1992), but such demand functions do not look at how the willingness to pay (WTP) for a visit is a function of the number of visitors. In terms of Figure 7.1, the WTP for a group of identical visitors OP, assuming that some critical number is not exceeded is given OB. The marginal cost per visit is OC. Each visitor will compare that marginal cost with the WTP as given by the line BZZ*. This results in a number of visitors equal to OV. However, the marginal visitor creates congestion effects on all other visitors, resulting in an additional or marginal value as depicted by the line ZZ**, which is below ZZ*. The socially optimal number of visitors is OW, but the free access equilibrium will result in a number equal to OV. The potential pool of visitors is OP.

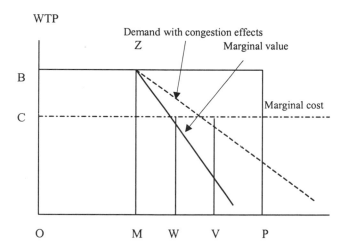

Figure 7.1. Congestion Effects of Tourism

Remarkably the literature on tourism does not contain serious estimates of this congestion effect. To be sure, there are estimates of the price demand elasticity of visits to sites using the travel cost method, but these estimates do not separate out the decline in the WTP due to the fact that people with a lower WTP are visiting the site (a factor we have eliminated in Figure 7.1), and the fact that the WTP of any one visitor declines with the number of visitors. If we are to develop tools for sustainable tourism it is precisely these kinds of data and analysis that are needed.

What kind of market structure is most appropriate to the tourism industry? This will vary according to the problem being analysed. One model is that of "open access" as presented in Figure 7.1. Individuals or small tour operators deliver visitors to a site, which is not controlled in any other respect with respect to numbers. This will apply to locations such as "free" beaches, pilgrimage sites, and some public monuments. Even with a site fee, to cover the local costs of maintaining the site, the problem of congestion is, of course, not addressed. The latter needs specific controls or market based instruments to address the congestion issue.

Costs, Prices

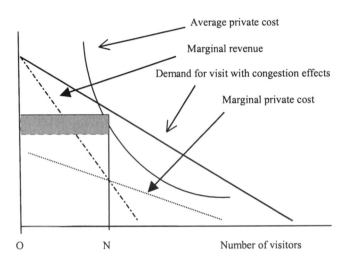

Figure 7.2. Private Optima with Monopoly (Profit is shaded area)

A second kind of model is one where a single operator, or a small number of competing operators, provide access to a site. Typically, in such cases there are significant "set up" costs and the industry is characterised by unit costs declining with the number of visitors, or increasing returns to scale. Such operator(s) will not take account of either the congestion dimension or of other external costs. However, by reducing output below the competitive level, they will make some move towards the social equilibrium. If there is only one operator, the industry will be a monopoly and the producer will maximise profits by equating marginal revenue to marginal cost. If the industry is open in the sense that anyone can provide the service, it will be characterised by monopolistic competition; a structure where each firm faces a downward sloping demand curve and maximises profits in the same way as a monopolist, but in the long run ends up making zero profits.

With monopolistic competition, some degree of product differentiation is expected. One way this could manifest itself is through "packages" in which different complementary products are offered, such as different grades of hotels, meals, guides. Another is through competing sites, such as ski resorts or beach resorts, varying in the degree of congestion. In the latter case, as will become clear later, some other aspect of quality variation is also required to fully determine the equilibrium.

In the case of controlled access equilibria and a monopolistic industry, numbers will be below those that would exist with a "competitive equilibrium", although the latter does not always make sense if the industry is one of increasing returns to scale. In any event, the monopolistic equilibria will deviate from the social optimum. It is not possible to say, in general, whether the industry will result in too few visitors or too many. Figure 7.2 below shows the situation for a monopolist. Equating marginal private costs to marginal revenues results in ON visitors.

But the social optimum will equate the marginal social costs to the marginal value as discussed above. The latter could be at a point greater or less than ON, depending on how the marginal costs compare with the marginal private costs. The same applies in the case of monopolistic competition.

3. SOME NUMERICAL SIMULATIONS

3.1. Congestion and open access

In this section we look at some simple models for visits to a site where congestion and external costs are present, and see how the market and social equilibria compare. The first case is that of congestion and open access. Assume that the demand for visits is form a uniform population, with each person willing to pay 100 euros for a site visit if the number of visitors does not exceed 500 on the day of the visit. Beyond that level, the WTP declines and the function is given by:

$$WTP = A - BN \quad N > 1000, \quad A > 0, B > 0 \quad (1)$$

The key question is how does the value of B relate to the cost of a visit? As we noted there are no real estimates of this congestion parameter. Consequently we have to make some assumptions. In Table 7.1 below we have taken "elasticities" of congestion ranging from -0.1 to -5. The potential number of visitors is taken as 10,000, all of whom have a WTP of 100 euros. When the number exceeds 500, a congestion effect comes into force. The elasticity of the congestion effects tells us by how many percent the WTP declines when there is a one percent increase in the number of visitors. Since the demand function is taken to be linear, the elasticity will vary with the number of visitors. The elasticities in Table 7.1 are calculated for WTP $= 100$ and $N = 500^3$.

The figures in Table 7.1 show that the more inelastic the WTP with respect to the number of visitors, the greater the difference between the social and private optima. Furthermore, the greater is the difference between the marginal cost of a visit and the WTP of the population, the greater will be the gap between the private and socially optimum number of visitors. Finally, the greater is the number of visitors at which the congestion starts to occur, the larger will be both the free access and social equilibria.

Table 7.1. Relationship between Private and Social Equilibria with Congestion Costs
(Figures given are number of visitors)

Elasticity of WTP with respect to N	Marginal Private Cost as Percentage of WTP							
	20		50		75		90	
	Free Access	Optimal	Free Access	Optimal	Free Access	Optimal	Free Access	Optimal
-0.10	4,500	2,500	3,000	1,750	1,750	1,125	1,000	750
-0.25	2,100	1,300	1,500	1,000	1,000	750	700	600
-0.50	1,300	900	1,000	750	750	625	600	550
-1.00	900	700	750	625	625	563	550	525
-2.00	700	600	625	563	563	531	525	513
-3.00	633	567	583	542	542	521	517	508
-4.00	600	550	563	531	531	516	513	506
-5.00	580	540	550	525	525	513	510	505

Notes: Potential number of visitors with WTP $= 100$ is 10,000; WTP declines when visitors exceed 500; WTP function is assumed linear; Elasticity is calculated for WTP $= 100$, $N = 500$

If we were to bring external costs into the picture, the divergence between the private and social optima would be even greater. Formally it is like raising the marginal private cost in Table 7.1; the higher the cost, the smaller a percentage of the total number of potential visitors is the number of socially optimal visitors. Incorporating the external cost does not change the free access solution, and so the gap between that solution and the optimum gets larger.

What kind of instrument would one use to bring about the social optimum? Recall that the problem arises because of the open access nature of the site being visited. It may be possible for the government to regulate visitors through a permit system, but that is often impracticable. Many sites have too many points of access to be able to control entry. For such cases checks on visitors at the site to ensure they have a permit may be required. In any event, if a licensing system is introduced, it is highly desirable to have licences subject to trade or auction. By that means those with the highest WTP can visit the site, thereby ensuring the greatest benefits. If the state wants to encourage students and others with low income to visit the site, it can issue free permits to such groups. If they then choose to sell them, however, it is difficult to rationalise why such sales should be banned[4].

An alternative system of control is to offer a franchise for the regulation of the site to one or more operators, who then have an incentive to internalise the congestion costs. Responsibility for access is then that of the operator and no fee is charged for the privilege. The government, may, however, need to impose a charge per visitor to cover the external costs, and/or a tax on profits to recoup some of the profits created by the franchise. The following section considers equilibrium with such a system of control.

3.2 Equilibria with Monopoly

Consider the following simple market structure. There is a single producer with a cost function

$$C = S + DN - EN^2 \qquad S > 0, D > 0, E > 0 \qquad (2)$$

The producer faces the demand function (1) given above. For simplicity, however, we assume that congestion sets from N=0. Simple profit maximisation yields a number of visitors of N_M of

$$N_M = (A-D)/2 (B-E) \qquad (3)$$

The second order conditions for a maximum require that E < B and an interior maximum requires that A > D.

Let us assume further that the optimal number of visitors under the monopoly is 1,000. It is not difficult to show that, at the profit maximising position, the elasticity of the demand function as given in the previous section, must be numerically less than one.

In order to present the simulations we need to give numerical values to a few more variables. These are as follows:

Elasticity of WTP with respect to N at N = 1000 and at P given by equation (1). We assume values of the elasticity of between -0.11 to -0.67.

Fixed or Set Up Costs (S). We assume that these range from 0.5 percent to 5 percent of the turnover of the monopolist.

Cost parameters (D) and (E). Given (3), A, B and N_M, there is only one degree of freedom to determine either D or E. Recall also that E < B. Hence the range of values for E are limited. We assume values of 0.4 B and 0.8 B.

External costs per visitor. We look for maximum external costs per visitor beyond which the socially optimal number of visitors is zero. This informs us as to what range of external costs can exist and still allow the site to be visited.

Table 7.2 below gives the solutions for the different values of the parameters. The model has been calibrated so that the optimal output for a monopolist is 1,000 visitors in each case. The corresponding socially optimal output is the same as the monopolist's output. The reason is that the monopolist is taking account of the marginal value from each visit, because that is also the marginal revenue to him. Hence the two equilibria are identical. This simple property will not hold, however, once we bring in visitors with different WTP for a visit. Then the monopolist will exploit the differences in WTP and produce too low a number of visitors.

Table 7.2. Output for a Monopolist and Socially Optimal Output

Elasticity of WTP with respect to price	Value of E	Monopolist output	Social optimal output with zero external cost	Value of external cost at which optimal N is zero
-0.11	0.004	1,000	1,000	12
-0.11	0.008	1,000 *	1,000 *	4
-0.25	0.008	1,000	1,000	24
-0.25	0.016	1,000	1,000	8
-0.43	0.012	1,000	1,000	36
-0.43	0.024	1,000	1,000	12
-0.67	0.016	1,000	1,000	48
-0.67	0.032	1,000	1,000	16

Legend: The value of A (the max WTP for a visit) is EURO 100 throughout; The value of B is given by $\eta = 1/\{(1/10B)-1\}$ where η is the price elasticity of WTP with respect to N.

Notes: * The output is 1000 only if the set up costs are less than 2.7 percent of total revenue. Otherwise it is zero.

The table provides some interesting insights. First the fact that the monopolist acts as a social maximiser is unusual. It arises, as noted above, because of the fact that the marginal revenue and the marginal value of an extra visitor are identical. Second, the range of external costs that can be accommodated within the model is rather small. For example, in the first scenario the maximum value of the external cost is only 12 percent of the WTP. It can rise to as much as 48 percent but it does depend on the precise values of the parameters B and E. Although linearity accentuates these switches, it is clear that the parameter values for the slopes of the marginal cost curve and the demand curve are critical. The policy implication is that the costs and the taxes have to be estimated with great care if the market based instrument is to get correct solution. There is little room for error in setting the tax. With other parameter values the scope for manoeuvre is a little greater but not that much. It should be noted that the upper limit of the external costs varies from 11 to 110 percent of the unit costs for a monopolist delivering 1,000 visitors.

3.3 Equilibria with monopolistic competition

In this section we consider the same demand and cost curves, but now open the control of the sites to competition. With the increasing returns to scale postulated in (2), a fully competitive equilibrium is not possible (when price equals marginal cost, the firm is making a loss). Hence a monopolistically competitive structure is required. Firm I has output N_I and all firms are treated as identical. Hence

$$N_I = N/F \qquad (4)$$

where F is the number of firms. The firms total revenue is given by TR_I, where

$$TR_I = (A - BN)N_I \qquad (5)$$

The conditions for profit maximisation and the requirement of zero profits yield values of F and N to be determined by the following equations:

$$F = \{BN - 2EN\}/\{A-D-BN\} \qquad (6)$$

$$AN - BN^2 - SF + DN - EN^2/F = 0 \qquad (7)$$

Solutions to (7) and (8) are given in Table 7.3 below. The most interesting features that emerge are the following:

• the number of firms emerging is highly variable, with many firms setting up when the set up costs are low and when the value of "E", the slope of the marginal cost curve, is small relative to "B", the slope of the "demand" curve;

- although the number of firms varies from 1 to 80, total number of visitors delivered is much more stable. It does not exceed 1,200 visitors, or 20 percent above the competitive output;
- the higher the elasticity of WTP with respect to N, the greater the number of firms and the greater the output.

Table 7.3. Equilibria with Monopolistic Competition

Elasticity of WTP with respect to price	Value of E	Cost of set up as percentage of annual revenue	Number of firms	Monopolistic competition	Monopolistic output
-0.11	0.004	5.0	1-2	1,050	1,000
-0.11	0.004	0.5	17-18	1,185	1,000
-0.11	0.008	5.0	0	0	0
-0.11	0.008	0.5	1	1,000	1,000
-0.25	0.008	5.0	3	1,125	1,000
-0.25	0.008	0.5	29	1,195	1,000
-0.25	0.016	5.0	1	1,000	1,000
-0.25	0.016	0.5	1	1,000	1,000
-0.43	0.012	5.0	5	1,155	1,000
-0.43	0.012	0.5	50	1,195	1,000
-0.43	0.024	5.0	1	1,000	1,000
-0.43	0.024	0.5	1	1,000	1,000
-0.67	0.016	5.0	8	1,175	1,000
-0.67	0.016	0.5	80	1,197	1,000
-0.67	0.016	5.0	1	40	1,000
-0.67	0.032	0.5	1	0	1,000

What are the implications of these results for the selection of instruments? First, in dealing with congestion, it is clear that the monopoly or monopolistic competition models can provide a good way of addressing the issue. In the case of the monopoly, some way of capturing the profits will be required. This can be achieved by a special tax on such profits, or by auctioning the right to operate a site. Prices will be higher than under competition, but the restrictions in numbers will be socially optimal. If the government wants to retain access to a site for low-income groups, it can do so through direct subsidies for them or through a differential pricing for such groups being part of the conditions of the monopoly. Second, the monopolistically competitive solution can bring about a solution that is quite close to the socially optimal one; it may be a reasonable price to pay for the benefits of increased differentiation that such competition offers. Third, where external costs are concerned, a tax or charge is justified. The magnitude of such a tax, however, will have to worked out carefully, as the solutions are quite sensitive to the precise values adopted.

4. FURTHER DEVELOPMENTS

The model presented so far has not looked at competition between sites and differences in the WTP between visitors. If there are several sites competing for the visitors, and not all are controlled by the same authority, there will be a disincentive to internalise external costs, on the grounds that you will lose visitors to the uncontrolled sites. It would be interesting to have some empirical estimates of how much substitution there is between sites (presumably as a function of distance) depending on the presence of taxes or other instruments. Such estimates are sadly lacking.

One way round the problem would be to have an agreement among countries within a region to impose charges based on external costs and not to compete on this issue. Within the EU, for example, such an agreement should be possible. Details of how the external costs are assessed will have to be worked out, but this is not an impossible task. Further international co-operation is more difficult but perhaps less necessary, given that the greatest competition is between sites that are close together.

Another way in which the model needs to be extended is by competing sites offering different services. It is possible, and casual empiricism supports this, for some sites to offer a "low cost/high congestion" package and others a "high cost/low congestion package". The above model of monopolistic competition is the most suitable for analysing such combinations. It is easy to bring in different WTPs for visits as a function of, for example, income and for sites to compete in terms of congestion levels. That alone, however, is not enough to characterise an equilibrium. Sites have to offer some other differentiating factors if the profit maximising and zero profits conditions are to be simultaneously satisfied. These could include different levels of hotel accommodation, restaurants etc. Work on such models is being undertaken by the present author. It suggests that, separating equilibria with different levels of congestion will emerge. Hence the congestion dimension can largely be handled through franchises, but the external costs will be different in different sites and differential charges will be needed, with higher charges in the more congested sites, which goes against any equity objective. Furthermore, if a site is to have many operators, a higher level decision will be needed to determine the overall level of congestion.

5. CONCLUSIONS

This paper has looked at the question of sustainable tourism in terms of its congestion and external cost impacts. With unregulated access to tourist sites, both of these result in levels of development that are unsustainable and socially sub-optimal. In order to determine the optimal numbers of visitors, however, we need information on the sources of sub-optimality that is quantitative. Much of this information is lacking.

The congestion issue can be addressed through a monopoly being provided control in the operation of a site. This will result in restrictions that are optimal with respect to the congestion dimension but will also generate high profits. Some way of clawing back these profits will be needed – through special taxes or through the auctioning of franchises.

In some cases an industrial structure based on monopoly is not feasible and monopolistic competition is more appropriate. This structure has been analysed and found to generate equilibria that do not deviate too much from the monopolists in terms of total numbers of visitors, but with a lower prices and a zero excess profit level. Hence such a structure may be acceptable as a second best solution, but this should be investigated further.

Internalising the external costs require data on the levels of these costs. Preliminary results suggest that this information needs to be rather precise as the solutions are sensitive to the chosen values of the charges.

Competition between sites has not been analysed in this paper, but is being studied. Preliminary results suggest that sites will offer different levels of congestion, as well as other quality differences, but that dealing with the external cost dimension will be more difficult. Some agreement to impose such costs will be required if the competitive model is to operate between sites under different jurisdictions. Furthermore the full implications of different external costs and therefore different charges need to be studies in greater detail. Indeed, the robustness of the results presented here to models that include: multiple visits, multiple stages in the delivery of tourist services, each with a different industrial structure, and visits with multiple objectives is essential. There is much work to be done before we can achieve a tourism industry that is fully consistent with the goals of sustainability. But it is clear that economic instruments have a role in that strategy, along with other methods of regulation.

NOTES

1. Italy, as the country with around 60 percent of the world heritage sites is the most vulnerable to this problem. It is entirely appropriate therefore, that this conference should held in Italy and that at least one paper should be devoted to looking at the regulation of visits to cultural monuments. For a general discussion of the conflicts between cultural protection and tourism, see Russo and van der Borg; and Fossati and Panella, this volume.
2. For a review of the literature on the use of sanctions and their impact on behaviour see, Cullis and Jones, 1997.
3. It is easy to show that the free access equilibria is given by $N(FA) = (A-MC)/B$ and the optimal position is given by $N(OPT) = A^*-MC/2B$, where MC is the marginal cost and A and B are as given in equation (1). A^* is the intercept of the marginal value curve going through $N = 500$ and $WTP = 100$.
4. For a review of tradable permits and their use in the tourism context see Renard, this volume.

REFERENCES

Cullis J. and Jones P. (1997), *Public Finance and Public Choice*, 2nd Edition, McGraw Hill, London.

Crouch G.I. and Shaw R.N. (1992), "International Tourism: A Meta-analytical Integration of Research Findings", in Johnson P. and Thomas B., *Choice and Demand in Tourism*, Mansell, London.

Fossati A. and Panella G. (2000), "Tourism and sustainable development: a theoretical framework", this volume.

Renard V. (2000), "Land markets and transfer of development rights: some examples in France, Italy, and the United States", this volume.

Russo A.P. and J. van der Borg (2000), "The strategic importance of the cultural sector for sustainable urban tourism", this volume.

Tietenberg T. (1997), *Environmental and Natural Resource Economics,* 3rd Edition, Harper Collins, New York.

8. LAND MARKETS AND TRANSFER OF DEVELOPMENT RIGHTS:
SOME EXAMPLES IN FRANCE, ITALY, AND THE UNITED STATES

VINCENT RENARD

1. INTRODUCTION

In comparison with other fields where tradable permits apply such as air or water, land has some quite distinct properties, not least because the many entangled legal instruments that govern it play a large part in determining its price. Property law varies widely from one country to another, particularly so in the case of development law, the prime cost determinant when land is in demand. For this reason, before going into the detail of actual examples of tradable permits, the relationship between property law, urban development regulations and tradable permits will be examined in Section I, with particular attention to the differences between legal systems of the type found in North America and those common in western Europe.

Following this, a description will be given of how the tradable permit system – which may still be deemed experimental since the practice is not yet widespread even in the United States – operates in three countries: the United States, France and New Zealand. From consideration of these experimental schemes, a number of conclusions will be drawn as to the value of the system, its effectiveness in reaching its goals and the conditions required for it to operate satisfactorily.

2. TRADABLE DEVELOPMENT PERMITS AND THE FRAMEWORK OF URBAN DEVELOPMENT

It is perhaps unnecessary to point out that the underlying reason is the economic efficiency generally expected from market mechanisms in contrast to the allotment mechanisms employed by administrative and planning authorities. Following the ground-breaking article by Ronald Coase (Coase, 1960) and subsequent theories of property law, it is generally accepted that definition and transfer of property rights, because they allow a market in rights to operate, give results that are "better" (in the sense of Pareto optimality) than the alternative of being unable to transfer such rights.

It should be noted that Coase's "theorem" rests on the basic premise that there are no (or negligible) transaction costs. However, the way property markets operate – and hence the way markets in development rights operate – generally entails high transaction costs and generates market imperfections (opaqueness, barriers to entry, oligopolies, monopsony situations, etc.) The underlying reasoning must therefore be applied with caution.

With regard to pollutant emissions, the introduction of tradable permits has the advantage both of allowing a measure of choice in ways to reduce pollution and of reducing the economic cost of complying with environmental constraints.

Their application to the implementation of urban development plans, to management of density in urban or peri-urban areas and possibly to areas of biological diversity not directly tradable (wetlands) still remains little explored, particularly in Europe.

The only significant area of use has been development rights, principally in the United States and to a more limited extent in France or Italy.

The introduction of tradable rights systems creates both conceptual and theoretical problems as well as practical difficulties in relation to implementation. In the case of the schemes considered in this paper, the technique would appear capable of reaching its goals only if it operates within a clearly defined legal framework and is closely tied to a rigorously applied land-use planning system.

2.1. Rights to pollute and land rights

Tradable rights have come to the fore as a topic in recent years, particularly in the general context of climate change, the greenhouse effect and air pollution, a contributing factor to the latter. An important threshold was crossed in late 1997 with the Protocol adopted in December by the

Kyoto Conference, which envisaged trading in quotas or emission credits. A primary conceptual difference should first be noted between a tradable permit attached to land and a tradable emission quota.

In the case of pollution, the object of the trade is an entitlement to emit an ongoing level of pollution, measured for example in tonnes of nitrogen dioxide discharged into the air per year. What is involved is a *continuing process* so that the relevant quotas themselves may go on being bought or sold *ad infinitum*.

The idea behind tradable land rights is quite different, since the right concerned is sold *outright* or *for a very long period*. Admittedly, it is only saleable in part, or may be bought back at a later date, but the purpose of the transaction is in no way to engage in an ongoing process. This will obviously have a major impact on the way the instrument is employed with respect to allotting rights and the conditions for buying them back.

Moreover, the term "tradable permit" is a concept widely applied in regulating pollutant emissions but is not readily transposable to land rights, or only partially so.

Any transposition there might be between pollution and development, or more generally between pollution and change of use (drainage of wetlands to plant maize, building a major road through woodland, urban development). Such a change of use, unlike pollutant emission, is a permanent right accorded by the regulations; it need not necessarily be used but if used can be used once only and in a manner that is often irreversible.

Instead of "tradable permits" it would be better to use the term "tradable rights" or "transferable rights". In the area of development or construction rights, the latter is the term used by the United States, which speak of the "transfer of development rights".

The concept therefore has to do with property law as applied to geographical space and reveals a major difference between th e legal systems originating in Roman law, based on the indivisibility and absolute nature of land ownership, and the diametrically opposite view of the main variants of Anglo-Saxon law, in particular North American law, which considers land ownership as a "bundle" of rights, some components of which can be treated separately, such as development rights, air rights or mineral rights.

Naturally, in the systems based on Roman law, legal concepts have also been introduced which allow one or more of these rights to be dealt with separately and even to be sold individually. Examples that come to mind are shooting rights, rights of way and mineral rights. However, these are distortions of the system rather than its underlying principle.

This difference in legal structures and in case law go a long way to explaining the difficulty countries of Western Europe or Japan have in coming to grips with concepts or techniques that are widespread in North America. It is particularly evident in the long-standing American practice of tradable environmental rights (easements), a concept which appears nowhere in French law for example, with its need to determine dominant tenement and servient tenement in order to establish such a relation.

2.2. The case with land: compensation of constraints

Central to the creation of a market in tradable rights is the issue of the financial and fiscal implications of urban development regulations.

To recap: in urban and peri-urban areas the value of a piece of land lies in the rights attached to it, which are conditioned by urban development regulations and by environmental regulations. Whenever regulations are introduced by public authorities, the price of land is strongly affected. Since any change to the regulations moves the price of land (up or down), it may be asked what corrective measures ought to be taken by the public authorities.

As far as urban and peri-urban areas are concerned (the problem being most acute in the latter), the response has differed from country to country. Roughly speaking, most countries in Western Europe have adopted the principle that constraints on urban development are not liable to compensation. As expressed in the French Urban Development Code for example, this principle applies to any constraint affecting the road system or prompted by health, aesthetic or any other considerations and concerned with such matters as land use, heights of buildings etc., or prohibition of development in given zones (article 160-5 of the Urban Development Code). A constraint on the right to make use of a given piece of land is not considered grounds for compensation unless it infringes a vested right (for example withdrawal of a building permit already granted) or a change in the previous status of the site resulting in direct, material and indisputable damage to property. This comes close to taking.

Strict application of this principle, which makes landowners subject to unequal treatment, has naturally met with considerable opposition and led to the generation of *de facto* and *de jure* loopholes.

In France for example, the introduction in 1976 of procedures for the transfer of development rights comes under this heading and was strongly attacked as a breach of the principle of the ineligibility of constraints for compensation. In essence it was the pliability of urban development regulations, for example with respect to land-use planning, that generated the inequalities stemming from this principle. Although the law is tradable to some extent, this process is far removed from the basic assumptions governing a perfect market.

The underlying principle is otherwise in the United States. Although the constitutional legality of zoning has been well established since the 1926 Euclid decision (village of Euclid v. Ambler Realty Co., Federal Supreme Court), the dividing line between admissible constraint under police power which is ineligible for compensation and excessive constraint eligible for compensation (taking) or expropriation (eminent domain) is constantly shifting. The term "taking" which may be read as "seizure" is a contemporary borrowing from the Fifth Amendment to the United States Constitution ("nor shall private property be *taken* for public use without compensation").

Although this provision, which has been in force since 1789, is closely related to the French Declaration of Human Rights, it is being applied in a quite different way in contemporary case law and its broad interpretation by the courts has led it to be considered in practice as the basis for litigation to test the legality of urban development regulations.

As illustrated by a number of recent decisions and the rise of the Movement of Private Property Rights, attention is drawn to recent developments which have tended to restrict the implementation of regulations and constraints (in particular the Clean Air Act and the Endangered Species Act) by systematically playing the compensation card.

The way the case law in the matter has developed over time matches the gradual change that has taken place in property rights law, in which a distinction is made between what is private property in the strict sense of the term (and may thus be put on the market) and what is common property.

The keenness of the debate in the United States, both from the purely legal standpoint and in the political arena, shows how crucial definition of the "bundle of rights" is.

2.3. Added and lost value

Postulating the existence of tradable rights assumes that there is something to trade, in other words that one of the parties is ready to relinquish an attribute of his property (the right to build for example) to another owner.

Whatever the circumstances, no market will operate unless the exercise is worthwhile, in other words unless there is a demand for rights. This raises the issue of initial allotment of rights. Here too there are two concepts to be distinguished depending on the methods used to value land and real estate, which are themselves based on the way property is conceived.

Under one concept, which is fairly widespread in northern Europe, ownership of land does not include a right to the value usually added by urban development or, in practical terms, by the installation of infrastructure, the cost of which is generally shared between the developers and the community. This is, for example, what happens in Sweden or the Netherlands (Needham et al·, 1995) but by means of different mechanisms (long-term land reserves for Swedish towns, quasi monopoly of towns in the development process in the Netherlands), most of the value added by urban development being collected by the community.

The other concept, which prevails for example in several countries in southern Europe (there is no "pure" case and it is therefore difficult to be more precise here) consists in allowing the original landowner to keep the capital gain subject to tax corrections – for example value added tax. In such systems, the introduction of urban development regulations or their amendment will generally be perceived as a constraint on previously held rights, the assumption being that ownership was at the outset unconditional and in particular included the development rights.

This is the context in which it is possible to conceive of trading in a "right" that is assumed to be in existence but the actual use of which has not been authorised. This point is essential to understanding the crucial importance of the original allotment of rights and the *contractual* nature of that allotment.

2.4. Unilateral regulations and tradable rights

One final concept needs to be clarified in relation to the central focus of this paper, the idea of a market in tradable rights. Many regulations applicable to geographical space are subject to various trade-offs between the public authorities responsible for permits and applicants. In an urban development project such as a planned-unit development this happens when the municipality and the developer negotiate what construction is permissible and what public amenities the developer will undertake to install.

It may also happen in the case of contractual land taxation systems in the United States where the contract negotiated covers (to some degree) both a commitment by the owner (for example to continue farming for ten years) and the basis on which his property tax is to be calculated.

These techniques – essentially fiscal in nature – whose aims are close (to make zoning constraints more acceptable to owners by offering compensation), use unilateral procedures such as assignment to a lower tax bracket or various methods of granting exemption from charges or according financial benefits in return for a commitment from the landowner (and/or user) to comply with given constraints, to maintain, to engage in nature conservancy, to continue farming, to engage in organic farming, etc. The literature dealing with such tax incentives is fairly extensive (see Cohen de Lara and Dron, 1997). They differ from the matters dealt with in this paper in that no market mechanisms are introduced; a unilateral decision is being made to offer a contractual tax technique. From this point of view, such contracts are outside the scope of this paper.

Other land rights markets exist. Mention may be made for example of the "market" in shooting rights, which is still extant although of minor importance (see Charlez in Falque and Massenet, 1997).

However, the heart of the problem and the major experiments in rights trading are concerned with the market in rights to "intensify" land use, which are essentially rights to engage in urban development and to build. These are the rights on which this paper will focus.

3. TRANSFER OF DEVELOPMENT RIGHTS: ORIGINS AND APPLICATIONS

This section will look at systems employing the tradable rights technique in operation in the United States, France and New Zealand.

The legal, economic and institutional background, however, varies widely in each case as do their aims and sphere of application. The programmes dealt with below were not chosen in response to a wish to achieve a statistically representative group – a difficult task – but from a desire to present a judicious mix of examples that are significant because of their scale, their duration, their management methods or their results.

3.1. United States

Origins

Section 2 discussed the connection between the case law on "taking", which became increasingly acrimonious in the 1960s and 1970s, and the advent of the technique of transferring development rights. In the face of stringent regulatory measures that involved the public authority in compensation, the technique helped to make the regulations acceptable while conserving public funds, since the value it added compensated for any value lost. This was the thinking that led to the emergence of transferable development rights in the early years of the century.

The city of New York was a pioneer in forging this technique, first by successive amendments to the 1961 Zoning Resolution, which made it possible to exchange densities between adjoining lots. The exchange of development rights among neighbours, or among a given block of houses, has been a widespread practice in developed countries and is generally compatible with the provisions of private law. A contract with the owner of a neighbouring piece of land provides a guarantee, for an agreed price, that it will be kept in good condition.

Termed "easements" in the United States or private law constraints (*servitudes de droit privé*) in Europe or in Japan, this form of exchange is generally limited to transactions affecting a small area and is only remotely akin to the concept of establishing a market in rights. The main idea is that of a right detachable from a given piece of land and usable in another area. A ground-breaking example in the United State is given by the New York Central Station, the case Grand Central Terminal v. City of New York having generated much written comment (Babcock and Siemon, 1985).

In this case, the city of New York was forced to provide compensation for the hardship caused by refusal of a building permit (the proposal was to erect a very tall 57-storey building above the station, which was in process of being designated a listed building) by granting the railroad company (Penn Central) development rights to a piece of land elsewhere. This was the first example of an out-of-zone transfer of rights, but it could not really be considered as the launch of a market in rights.

It was the 1970s that saw the real emergence of tradable development rights as a technique, one mainly applied to large areas of natural scenery.

The number of noteworthy examples is in fact quite small. Although much has been written on the subject, its actual impact has been limited; throughout the country, a little over 20,000 hectares have been protected by means of transfer of development rights.Table 8.1 shows the main features of four of the most noteworthy examples.

Pinelands National Preserve (New Jersey, US)

One of the most important examples of application of the tradable development rights method (transfer of development rights) has been progressively implemented over the past twenty years in this area covering almost 4,000 km^2 in southern New Jersey, one of the most populous regions of the United States, situated between Philadelphia, New York and Atlantic City. Lying close to a very large groundwater basin, it is also extensively wooded and home to many animal species.

Table 8.1. Programme description

	Pinelands	Montgomery County	Lake Tahoe Basin	Santa Monica Mountains
Population	450,000	781,022	51,775	11,336 (Malibu)
Year Program Initiated	1980	1980	1987	1979
Goals	Cohansey Aquifer and Forest Conservation	Agricultural Land Preservation	Watershed Preservation and Recreation	Erosion and Water Quality Control, Avoid Public Service Strains
Zoning or Permit-Based	Zoning	Zoning	Permit	Permit
Mandatory or Voluntary	Mandatory	Mandatory	Voluntary	Voluntary
Dual or Single Zone	Dual	Dual	Single	Single
Number of TDR/TDC Transactions	1,424 Severed Rights	4,300 Transfers	372 Coverage Transfers	493.7 Transfers
Land Area Preserved	12,538 Acres Preserved	Approx.34,000 Acres Preserved	Coverage Transferred CA: 66,686 sq.ft. NEV: 9,361 sq.ft.	864 Lot Restricted
Current TDR/TDC Value	$3,500-4,500	$10,000-12,000	CA: $5-8/ sq.ft. NEV: $8.5-15/sq.ft.	$17,000-21,000
Processing Period	30-45 Days	4 Days	2-11 Weeks	2-4 Weeks

Source: Johnston and Madison (1997)

It was with the express intention of protecting the environment that the United States Congress decided in 1978 to establish the Pinelands National Preserve and requested the Governor of the State of New Jersey to establish a special planning commission and to impose a moratorium on development in the zone while awaiting adoption of a development law. The plan ultimately adopted in this context, the Comprehensive Management Plan, was approved in November 1980.

The exceptional nature of the conservancy policy applied in this remarkable area is worthy of note. Preparation of the plan came first, the transfer of development rights was merely an instrument designed to apply it. It involved very stringent zoning, in which areas of land (and bodies of water) were divided into 10 categories, some zones being designated as closed to development but able to transmit development rights and others being classified as receiver zones where development was permissible subject to the purchase of rights from landowners in transmitter zones.

The exact mechanism is as follows: the allotment accorded each piece of land is determined on the basis of several criteria (location, type of terrain, past and present use, existing buildings if any, and method of land acquisition). Agricultural land and areas for conservation because of their outstanding natural beauty, the interest of their flora and fauna and the need to protect the aquifer (an extremely large one, over 60 thousand million cubic metres) are classified as transmitter zones, which are allotted development credits on scale which is subject to change. Four credits are required for a permit to build a house, i.e. to receive one development credit or DC.

In the core conservancy area (18,000 ha) where development is very restricted, one development credit is allotted for every 16 ha, one for every 8 ha in an agricultural zone and one for every 32 ha in a wetland zone. These basic parameters are adjusted to take account of various factors.

The receiver zones, which generally lie on the periphery, cover an area of 32,000 ha, and have a development potential of 126,000 constructions, or approximately twice that available from the transmitter zones.

The initial allotment process was clearly a key procedure and was carried out by the New Jersey Pinelands Commission. A threshold price of US$10,000 per credit was originally set, but the first exchanges rapidly led to a much higher market price. The sales price of a credit in 1996 was in the region of US$20,000, the credits received by landowners in the farming zone were thus worth FF 1/m². Once purged of its development credits, the transmitter land was closed to development in perpetuity.

Since 1980 the system has been managed by two bodies, the Pinelands Development Credit Bank (or Pinelands Bank) for the state and, for the main county involved, the Burlington Exchange. The latter buys in credits from Burlington County only but may resell them anywhere in the zone.

During the period 1981-1996, only 6,000 ha of the 65,000 ha in line for closure for development were in fact so closed and 4,000 of the planned 26,000 houses have been built.

Quantitatively speaking the original aims of the programme have by no means been met in practice. That said, the programme is a long-term one and the finite upper limit placed on the number of development credits justifies the length of the process.

This is the most extensive and most developed example to date. The tradable rights programme has played its part there, albeit a minor one, not least as an educational tool to promote acceptance of the remarkable planning effort made by the Pinelands Commission.

It has also been accompanied by a heavy emphasis on public ownership of biologically rich zones. So far these make up 170,000 ha and will ultimately cover 200,000 ha.

Montgomery County (Maryland)

This, like the Pinelands Preserve, is a noteworthy and interesting example because of its size and the length of time it has been in operation.

Located in north-west Maryland, close to Washington D.C. and Virginia, it is a county facing the usual problems of protecting areas of outstanding natural beauty from strong urban development pressures.

Despite relatively rigorous urban planning regulations in place since the early 1970s, in particular the requirement for a lot to be at least 2 ha for building to be permitted, nearly 20 per cent of the county's agricultural land was swallowed up between 1970 and 1980. This led the County Council in 1980 to draw up a programme for transfer of development rights as part of the machinery for implementing the Functional Master Plan for the Preservation of Agricultural and Rural Open Space. The special commission responsible for execution of the programme (the Maryland National Parks and Planning Commission) decided to establish a rights-transmitting agricultural zone occupying 40,000 ha of prime agricultural land, where the permissible housing density was reduced from one dwelling per 2 ha to one dwelling per 10 ha – the difference between the two constituting the tradable rights.

Only a few receiver zones had been designated when the project began, but they were not officially recognised as such until later; preference was given to zones with infrastructure already in place, in order to prevent allotment of development rights from being too much influenced by existing expectations. According to the players involved (see Canavan, 1993), the designation of the receiver zones was the crucial step together with allotting a development right that gave sufficient incentive to set the scheme in motion.

The procedure was simpler than in the Pinelands case, residential development only being authorised. The ratio for a transmitter zone is one development right per 2 ha, no splitting being allowed, regardless of the nature of the terrain. Final severance of development was entered on the title deeds by the county administrative authorities.

The county has established a bank (the Development Rights Fund) that buys and sells development rights and may provide security for loans backed by tradable development rights.

From a starting price of US$3,000, the cost of a development right subsequently rose to between US$10,000 and US$20,000. The value of the right in the transmitter zone is thus US$0.5/m^2.

The system has worked well. By the end of 1997, 4,300 of the 4,700 rights had been transferred, giving permanent protection to almost 15,000 ha. The success of the scheme became apparent when the conversion of agricultural land slowed down significantly. Although the conversion rate had been 1,000 ha a year before 1980, that was approximately the total area converted between 1980 and 1991.

A high pressure information campaign and the simplicity of the procedures involved were deciding factors in Montgomery County's success.

In addition, the procedure clearly made the development restrictions introduced by the plan more easily accepted. The price of the transferable development right (equivalent to FF 1/m^2 in the farming zone) was not a decisive factor in this process. The price has in any event been sluggish for the past two years. However, owing in particular to the active and participatory attitude of the Park Commission, it has helped to make rigorous zoning more acceptable.

Lake Tahoe (California-Nevada)

This is undoubtedly the best known example, not only because of its land management method, but also because the shores of this natural lake of 80,000 ha, of crystalline water are outstandingly beautiful and naturally much sought after, on the line between the states of Nevada and California.

The zone as a whole (the lake straddles the border between the two states) is managed by a special agency, the Tahoe Regional Planning Agency (TRPA), which is responsible for planning and for issuing development permits. Under conflicting pressures from owners anxious to increase the value of their land and residents hoping to limit development, and in addition faced with water resource constraints, the TRPA began in the mid-1980s to make use of a system of tradable rights with very special features linked to the water resource.

The tradable right here is the coverage coefficient, that is the portion of the surface area of a lot that may be built on. Calculation of the transferable rights allotted to each lot is determined on the basis of a calculation of ecological sensitivity producing a set of 8 factors connected to type of soil, slope, hydrological features, exposure, etc. This sophisticated lot evaluation system (IPES, Individual Parcel Evaluation System) was established after lengthy public debate. The evaluation process is carried out by a team of three (a soil specialist, a hydrologist and a town planning expert) and ends by assigning a factor to each lot from which the coverage coefficient can be derived. Since the introduction of the evaluation system in 1989, over 12,000 lots have already been assessed in this way.

From coverage transfers it can be seen that prices are higher in Nevada (US\$85 to 150/m^2) than in California (US\$50 to 80/m^2). Although the system is quite complex it is now reasonably well practised, evaluation taking six to eight weeks and the transfer itself two to three weeks.

The market in transferable coverage is managed in California by a special body, the California Tahoe Conservancy. The Land Coverage Bank provides a cash reserve for purchase and restoration of degraded sites as well as for purchase and sale of land coverage rights. By late 1995, almost 400 transfers had been made.

The Lake Tahoe project faces a special problem. Since the limiting factor is the water resource, the total rights now allotted represent more or less the ultimate limit to lakeside development. It is therefore conceivable that the market will come under pressure as that limit approaches. Should prices then be allowed to rise to levels unforeseen today? (Prices are already at US\$200/m² on the Nevada side).

3.2. France

As indicated in Section 2, French urban development law – as in most European legal systems – is based on the principle of no compensation for urban development constraints or, in the terminology used in North America, "police power" takes over from "eminent domain". This feature clearly creates inequities for landowners whose land is differently affected by urban development schemes.

In the French case, such inequities posed a particularly thorny problem in the early 1970s with the launch of urban planning by the Land Use Act (Loi d'Orientation Foncière) of 31 December 1967, which for the first time introduced zoning as a general principle (Plan d'Occupation des Sols) from which there was no exemption. After a number of abortive attempts and lengthy controversy, the Urban Development Reform Act (loi sur la rèforme de l'urbanisme) of 31 December 1976 finally made it possible, in some zones, for development rights to be transferred from one subzone (transmitter) to another (receiver). This principle is today embodied in article L 123-2 of the Urban Development Code:

"In zones to be conserved because of the quality of the landscape ... land-use plans may determine the conditions under which the development potential determined by the land-use coefficient set for the zone as a whole may, subject to authorisation by the administrative authorities, be transferred in order to promote concentration of development on other lots in one or more sectors of the same zone."

This wording is somewhat ambiguous in its reference to zones "to be conserved because of the quality of the landscape". Is the intent to exclude productive agricultural areas as well as zones available for development? Can such a distinction be made? There is no case law as yet on this point.

The provision generated a great deal of impassioned debate before, during and after its adoption. Considerable criticism has been levelled on three points: the tie to zoning, incompatibility with the principle of no compensation for constraints, and distributive justice.

With regard to the first point, the difficulty resides in application of the term "quality" to landscape which seems to equate it with a conservancy area, in which all development is prohibited. But how can productive agricultural areas be excluded since they are often areas in which the landscape is of outstanding quality?

The pressures that could be brought to bear on owners of agricultural land to apply for transferable development rights can be imagined. This ambiguity in the relationship to zoning has had an adverse impact on the first trials of the technique in France, for example at Lourmarin (Vaucluse).

On the second point, what is being called into question is the very principle of transfer of a development right, in that it implies compensation for constraints, since it applies solely to land to be "conserved because of the quality of the landscape".

From the point of view of distributive justice, it may be said that the equity of the procedure is basically conditioned by the original distribution of land holdings. Where land has been distributed in a fairly equitable way among the inhabitants, the procedure would be neutral in terms of distributive justice.

However, the situation is generally very far from this. The technique then represents a transfer from the community as a whole, the legitimate beneficiaries of any value added by urban development, to the subgroup represented by the landowners in the zone concerned.

Some example

Although the law allowing transfer of the land-use coefficient has now been in place for 23 years, examples of the practice remain few, disparate and difficult to summarise.

To start with, the examples are confined to a limited number of geographical areas.

Details are given below of five of the earliest examples, located in the French Alps and in the south of France.

Taninges – the Praz de Lys plateau

In this relatively clear-cut case, the task was to manage and organise the development of a gently undulating plateau at an altitude of between 1,600 and 2,000 m, where the expansion of winter sports, in particular cross-country skiing, was to lead to the building of a road to make the plateau accessible to motor traffic.

To prevent haphazard development, a development plan was drawn up before the road was built which divided the plateau as a whole (zone to be conserved because of the quality of the landscape) into development rights transmitter zones (in general accorded a transfer right of 0.035) and receiver zones where development was permissible, with development rights varying between 0.10 and 0.30, whose owners were permitted to build provided they had acquired additional rights from owners in the transmitter zones. It should be noted that the commune, which owned land in a transmitter zone, was granted a COS (floor-area ratio) of 0.10.

To give an example, the owner of a 7,000 m2 piece of land in a receiver zone of density 0.20 has, in terms of his entitlement of 0.035, a surface area available for development of 7,000 × 0.035 or 245 "points". In order to build at the authorised density of 0.20 he needs 7,000 × 0.2 or 1,400 points, and must thus buy 1,155 points, which must come from 1,155/0.035 or 33,000 m2 of land that as a result will become permanently closed to development.

The system has worked fairly well. It should be noted that the commune is heavily involved in its operation and initiated the movement itself by making the initial transfer between two lots belonging to the commune, one in the transmitter zone and the other in the receiver zone. The private market then took over, leading to prices of between FF 300 and FF 500/m2 as the floor price before work began. The compensation paid to owners in the transmitter zone, whose land then became closed to development in perpetuity thus came to FF 10 - 17 /m2.

Although not negligible, this sum represents only a fraction of the value of building land in the zone. Furthermore, the value has remained unchanged or even dropped over the past four years.

Lourmarin (Vaucluse)

The situation in the commune of Lourmarin is quite different in its origins, practice and the conclusions to which it leads. It is an attractive area in the Vaucluse where land is already being "nibbled away" by both primary and second homes.

The process, launched in the late 1970s by a resolute municipality, included agricultural areas in transmitter zones. This set off a long-standing dispute with the urban development authorities, which have remained unenthusiastic about the technique. It has however worked well under the strong leadership of the municipality which launched it and maintains the "market" in development rights.

A development rights "exchange" has been established, run by the mayor's office and the Loumarin notary. Currently about three-quarters of the housing density foreseen for the receiver zone has been used. However, the price for transferable rights is remaining relatively steady at FF 15,000 to 20,000 per 50 m^2 of land where building is permitted. Since the density allocated to the transmitter zone is 50 m^2 per hectare, development rights thus cost FF 1.5 to 2 per m^2 of building land. This by no means fills the gap between the price of agricultural land and that of building land, but, as the mayor says, has given a breath of air to landowners in the conservancy area closed to development. In view of the complexities of managing the system and its modest results, the municipality does not intend to take the scheme further.

Les Gets

This is another example of what can be done by a small mountain commune, an effort that has been underway for 20 years, under conditions the commune considers to be satisfactory. The urban development plan has designated a fairly large number of transmitter zones, and allotted a development right equivalent to a density of 0.05, the development rights ratio being 1/3 from the receiver zone and 2/3 from the transmitter zone.

The process was slow to take off. Often a transfer led to no outcome (building project abandoned) and the number of transactions remains small (10 to 20 a year) although representing nearly 50 per cent of the entire market. The "price" of development rights, after a fairly rapid rise (to FF 50 per m^2 of land in the transmitter zone), has tended to stagnate or even fall in the past few years. In any event, the level is low in relation to the price of building land (FF 250 to 400 per m^2).

Two comments are called for by the Les Gets experiment. Firstly, the process initially had difficulties in matching supply to demand in development rights. A pragmatic solution was found in that the notaries, who in France are "ministerial officers" or in other words public officials, agreed to take on the task of organising the development rights market. As mandatory go-betweens in all transactions, they circulated information between those offering development rights for sale and those seeking to buy them. They played a key role in the operation of the scheme.

The second notable feature has been the way the mechanism is perceived by landowners. The practice is a long-standing one (20 years old) and has accustomed owners of land closed to development to think of their land as having a development right attached to it, whatever the final location of the development. It therefore introduced a stiffness, a sort of ratchet effect, into the zoning procedure.

Lastly, the commune introduced an incentive making it mandatory for development rights to be acquired in the largest, and most attractive, of the zones set aside for urban development. This procedure – which is generally found satisfactory – illustrates the importance of the connection between zoning and the transfer of rights technique.

Since only a small fraction of the available transferable rights have yet been used, the reallotment of new rights is not yet a problem.

3.3. Italy

The case in Italy is clearly different inasmuch as the basic legislation is quite different from what it is both in France and the United States, but it can also be underlined that the method has received serious implementation, especially during the last decade. However, the application is not totally convincing, be it in terms of efficiency in land allocation or in terms of equity.

The very principle of separating development rights from the general right of property has been introduced through a 1977 statute ("legge Bucalossi") that has instituted the "concessione di edificare", i.e. stated the principle that, beyond a very limited amount (0.03), the density allowed by a building permit should be "bought" by the beneficiary of the permit to reach the authorized density.

The application of the statute has not been very easy in urban areas, but the basic principle remained about the possible separation of development rights and their being traded. The effective application of the mechanism was quite limited until the late 1980s, but some interesting cases have been developed in the more recent period. Table 8.2 illustrates the type of uses and density in six Italian cities.

The two cities of Torino and Ravenna will be studied here more carefully[1].

Torino

The technique of transfer of development rights applies on various zones in the city of Torino, with a special focus on derelict zones that are supposed to be transformed and reused. The main feature of the program is the densification of some subzones – and the rehabilitation of others as open space – through a powerful incentive in terms of authorized densities. The method can thus be parallelled to bonus zoning in the United States. The previous table indicates the range of densities that apply, from 0.70 to 0.03.

*Table 8.2.*Uses and density in six italian cities.

City	Generalized development rights	Type of area	Authorized density
Casalecchio del Reno	yes	derelict urban land	0.23
		peripheral areas	0.12
Reggio Emilia	yes	fallow land < 3 ha	0.40
		transformed areas	0.25
		green spaces	0.10
Plaisancia	yes	fallow land < 3 ha	0.50
		fallow land > 3 ha	0.35
		industrial areas	0.30
		multifunction areas	0.30
		military zones	0.25
Ravenna	no	greenbelt	0.10
Torino	yes	urban areas to be redeveloped; service areas; urban parks; natural areas	0.70
			0.23
			0.05
			0.03
Parma	yes	inside urban areas	0.50
		outside urban areas	0.15
		conservation urban arcas	0.25

This program has been applied for the past three years, and it looks rather successful since, up to now, 2.5 million m2 have been or are being transformed, out of a total of 5.6 million m2. Paradoxically, the management of the program has been specially delicate in areas with a (possible) high density. In fact, this did not correspond precisely to market demand, in terms of density as well as in terms of economic activity (crisis of tertiary activities).

A second feature of the Torino program is that it has not really boosted the development of a market of tradable rights. Practically, transactions on development rights have occured simultaneously with the transaction on land.

Ravenna

This program applies since 1993 and was part of the development plan of the city. The use of TDR is strictly restricted to two key objectives: the protection of the greenbelt and the redevelopment of the harbour. A low level of development rights – 0.1 m2 / m2 has been granted to lands included in the green belt, even 0.033 m2 / m2 in areas close to rail tracks or roads. These DR can be transferred only in the redevelopment of the harbour. On the other side, landowners in the harbour area receive a density premium if they buy transferred DR.

After a lengthy beginning, the program is now being implemented in good conditions, boosted by the municipality through efforts to encourage such transactions. It can be noticed that some owners in the development area (the harbour) have bought land (freehold) in the greenbelt and kept it as a reserve of DRs.

As a conclusion, it can be commented that the application of TDR remains limited in Italy. Two further comments may be added.

First, the application of TDR must be explicitly articulated to the planning mechanism and its relevance guaranteed in terms of economic demand. In the case of Torino for example, the incentive was not strong enough to move the markets.

A second difficulty arises from transaction costs. Several transactions had to be abandoned because of the complexity and time length of negotiating the purchase of DR with a series of landowners, often reluctant as soon as the amount appears quite limited. In fact, the case of Italy makes it clear that, beyond political will and technical skills, its efficiency relies on a sound and articulated application of traditional tools of city planning.

4. SOME CRITERIA FOR AN EVALUATION

Trials of the practice have not yet reached a critical mass that would allow statistically reliable conclusions to be drawn. Even though there are a fairly large number of examples in the United States, they are in different geographical areas, have different aims, use different operating methods and, naturally, show different results.

Many examples of the practice – generally on an informal level – occur among very small groups of owners and operate by consensus without any formal legal or institutional framework being involved. Such examples, generally occurring in built-up areas, have a long history in the form of TDRs in the United States and private law constraints in France.

On the other hand, use of the method in a vast geographical area by means of a universally applicable mechanism formally established in advance is still fairly limited and represented by a few widely quoted landmark examples such as the Pinelands Preserve, Montgomery County or Lake Tahoe in the United States. However, most have features specific to themselves that make it difficult to reach general conclusions.

The examples presented and the information available on other schemes do nevertheless enable some assessment of the practice of attaching tradable rights to land to be made with respect to the various aims it hopes to achieve. After an examination of the real nature of the rights being traded, a look will be taken at the key link between tradable rights and zoning.

In addition, a consideration of the way these rights markets operate and the prices (market prices?) they generate will illustrate the distributive effects of the system.

As already mentioned, experience of the technique is still limited and difficult to interpret in general terms. Although as a practice it is fairly widespread on a very small scale in urban areas for the purpose of sharing development rights among neighbouring owners, relatively large-scale examples where a true market can operate and transparency and atomicity generally prevail are much less common. Such examples are of considerable interest and have produced substantial results, but each has its own story and has only come into being through a combination of factors (economic, legal, institutional and development factors) that are not easy to reproduce. This explains in part the anecdotal nature of the examples described.

4.1. Explicit objective(s)

The aims attributed to most schemes are generally environmental or architectural. The most frequent is nature conservancy (Pinelands, Lake Tahoe, Taninges).

Many examples, often smaller in size, focus on architectural merit and the preservation of listed structures. This is generally the case in New Zealand, in particular the major scheme in the city of Auckland. Nevertheless, the most frequent case remains nature conservancy, preservation of sites of outstanding natural beauty and protection of agricultural land in the vicinity of built-up areas.

Indeed, it is often in terms of surface area preserved for conservation in perpetuity that the success of a programme is measured, an area conserved in perpetuity being taken to mean one that has transmitted all its development rights and is thus closed to development. However, this particular aim is frequently merely the backdrop to the prime objective of the technique which is distribution, namely to provide compensation for the constraints society places on the use of property, or in other words to render acceptable the inequalities created by zoning laws, which by their nature cause development rights to be distributed inequitably.

The goal of nature conservancy, or of the preservation of structures of architectural merit, is the prime objective of the regulatory procedure. The transferable development rights technique is therefore more to be regarded as an intermediate instrument to facilitate implementation of a plan. Section 4.6 will discuss in more detail the concept of equity promoted by transferable development rights schemes.

The technique may also serve as a legal safety net for the planner. Even if the scheme is not in operation, the mere fact that it is in place will enable disputes over compensation for constraints to be avoided. This is an important aim in North America in view of the widespread litigation prompted by "taking".

4.2. The legal status of transferable rights

In all countries that have made use of transferable development rights of one sort or another, the legal status of those rights has been a point of contention. Are they an integral part of property (even when destined to be used at another site) or are they merely a financial instrument to provide compensation for value lost as a result of a constraint?

This is an important point, both because of its impact on the legal appreciation of the issue (in France or Germany for example, it is unlawful for planning restrictions to be liable to compensation except in very special circumstances) but also because of the way it is applied and in the way compensation is assessed.

As explained earlier, in the United States and Canada the case law on zoning ("taking") has played a key part in determining the way urban development operates and development rights are managed by providing for an owner to be compensated when a regulation considered extremely restrictive is viewed as "taking". With the sword of Damocles represented by the threat of litigation under the Fifth Amendment to the Constitution hanging constantly over the planner, he may find transferable development rights a means of mitigating the threat.

It was in fact as a consequence of a proceedings of this sort (the decision in Penn Central Transportation Company v. City of New York, 1978) that the practice was introduced. Case law, however, while it may have assimilated the principle that the title to one plot of land might include, even if only implicitly, a right to make use of another plot, has set limits to the practice, which however have not yet been finalised.

The concept of property rights itself has never been finally defined. Many commentators turn to the idea of a "bundle of rights", whereby ownership of land is constituted by a series of autonomous, separable rights – rights to use, to develop, to overfly, to cross, etc. However, this idea does not settle the question of transferable rights in legal terms, namely which of the rights attached to land are by their nature part of ownership (such as the right to farm the land) and which are rights whose attribution may be determined by the social function of the property (such as the right to build etc.).

The Penn Central decision, which had a major impact on the case law on "taking" in the United States, according to Judge Brennan, expressly took the stand that the transferable development right was a useful instrument to secure compensation for some of the effects regulation had on land and housing values. In Judge Brennan's own words explaining the majority view, "while these rights may not have constituted just compensation if a 'taking' had occurred, these rights nevertheless undoubtedly mitigate whatever financial burdens the law has imposed on appellants and, for that reason, are to be taken into account in considering the impact of regulation". Whether or not "taking" had occurred, the procedure was an additional instrument that could help landowners to accept stringent zoning regulations.

The recent Suitum decision (Suitum v. Tahoe Regional Planning Agency, 1175 ct. 1659. 1997) gives a good illustration of current trends in the relevant case law, which has taken a clear stand on the point. Under the Lake Tahoe transferable developments rights scheme described earlier, the appellant, who had been refused a building permit because his land was listed as a Stream Environment Zone was awarded tradable development rights to compensate for the loss of value of his property. He then sued the TDR scheme on the grounds that the right to build was an inalienable part of property rights, was as such usable *in situ* and was not transferable to another site. This was the first time such reasoning had been taken all the way to the Federal Supreme Court.

The Suitum decision was thus of major importance because it questioned whether it was lawful for the procedure to set limits to the financial implications of land-use regulations.

Such a development could well put a real brake on the practice of awarding compensation in the form of TDRs and thus strengthen the powerful Movement of Private Property Rights. The courts could thus find themselves facing an alternative: either "taking" occurred and is liable to full compensation, or it did not occur and there is no liability for compensation.

The decision did not in point of fact categorically rule out the technique as a means of providing compensation, but drew attention to the defects and inconsistencies of the Lake Tahoe scheme.

In a quite different context, a similar debate has been going on in France, although from a different starting point since the basic premise is that constraints are not liable to compensation. When the Act was drafted, voices were heard denouncing the risks involved in introducing rights that could be considered as imaginary (Lenotre-Villecoin, 1975).

Like the Suitum decision in the United States, this amounts to an attack on the very principle behind the creation of the legal entity of "the transferable right". According to J. Lenotre-Villecoin, the capacity to transfer an imaginary development right establishes a *jus abutendi*, or a right of disposal, in a case in which the public interest, in the form of regulation of urban development, is against existence of the right to build at all. Perhaps it is time to deal finally with the inseparable trio (*usus, fructus* and *abusus*) which "abuse" opinion in Latin countries.

4.3. Zoning and transferable rights

There is a clear link between the two. The transferable rights procedure is in itself a zoning instrument since it implies a division into transmitter and receiver zones. Greater precision may well be introduced by stipulating that the zones must be of precisely specified dimensions otherwise the whole scheme will be invalidated. If the scheme is to operate properly, owners in both transmitter and receiver zones need to be given the right incentives, which should help to balance supply against demand with respect to development rights.

In the case of receiver zones, where conventional planning regulations operate as usual, the purpose of zoning is to ensure a high standard of urban development. The quality of urban development is thus the criterion to be taken into account.

As the process proceeds, however, it becomes difficult to provide adequate incentives. Many schemes, such as the Pinelands one, use a system of "bonus zoning", in other words the density authorised increases if transferable rights are purchased. This makes it very tempting for the planner to reduce the "ordinary" density (where no rights have been purchased) and increase the "bonus" density. However, such a policy is likely to fall foul of the principle of vested rights and to lead to litigation. Incentive zoning is thus a difficult process to handle

Another sensitive issue is the eligibility of a zone to be designated a transmitter zone, which opens the way to a grant of transferable rights. The subject is one of endless debate with no evident way of settling it on a systematic basis

First of all, it involves agricultural land. Should such land be allocated transferable development rights and if so on whose behalf? Generally speaking, the price of agricultural land could be considered to reflect its productivity, the current net value of its future yield. It is paradoxical to allot "development rights" to land on which farming is expected to continue even if the rights are not to be used on that land.

Arguments are often based – this is generally the case in France and in most of the American examples – on the natural beauty and biological diversity of the site. However, this takes no account of existing usage and places owners having quite different relationships with their land on the same footing.

Although no general conclusion can be reached on the linkage between zoning and transferable rights, note should be taken of the risk of distorting zone demarcations and urban planning regulations in order to make the rights market work. It is important to maintain a proper perspective; the transferable rights procedure is no more than an aid to good urban planning and not an end in itself.

4.4. Technical and institutional feasibility

The examples described and the above comments make it clear that sensitive management is needed to make the technique work.

At institutional level, it is important to do away with the misconception that the transferable development rights procedure requires strong and stable institutions if it is to operate properly. The landmark examples in the United States all rely on an ad hoc institutional base – such as the Pinelands Commission – which assumes responsibility for the whole area concerned, without any right of oversight by common law institutions, counties or districts. This structure takes care of both zoning and management of development rights.

On the other hand, when an elected council is responsible for operation of the system, an election often succeeds in undoing a previous body's achievements. Another difficult matter, in addition to zone demarcation as mentioned above, is management over time. The initial allotment of rights in transmitter and receiver zones has to correspond to some point of equilibrium at some given time in order to enable the market to operate, and give sellers an incentive to sell and buyers to buy.

However, what should the long-term dynamics of the system be expected to be? From the standpoint of the seller a plot of land that has been stripped of its development rights is closed to development in perpetuity. Then as time passes and if demand remains at a given level or increases, pressures will build and the price of rights will soar. Should more rights then be allotted to land that has been stripped of them and under what circumstances?

Can the initial allocation of development rights be considered immutable (this is the view of the Pinelands and Lake Tahoe schemes) even if this is likely to lead to a price boom? What should be done with regard to owners in the transmitter zone who hold on to rights, speculating on a rising market? Should their development rights be expropriated?

None of the examples discussed has yet had to face this problem. Either the mass of rights has been very large, or demand has been overestimated, or incentives were not strong enough, or the scheme was too new, or the problem has not yet taken an acute turn. Its advent is inevitable and the reallotment of rights will require very careful handling. In any event, schemes need at the outset to determine what the medium-term prospects are and to define the reallotment procedure to be used.

Another technical question to be resolved is the way rights are to be managed. In order to allow the market to operate there has to be a body with the power to buy, stock and sell rights at any time. In view of the special nature of such a market and the likelihood of discontinuities, it is a mistake to think that it will be able to operate on a decentralised basis from the start.

Here again, the examples that are working well are doing so with the help of a "development rights bank" which operates in close association with the general aims of the system.

One last need is to educate owners, both transmitters and receivers. Trading in transferable development rights is an activity that is not self-evident and needs extensive explanation and demonstration in order to work properly.

4.5. Is the price of rights a real market price?

It would be pleasant to be able to answer this question, taking market price to mean the price that would balance supply against demand under conditions of atomicity, transparency, etc.

Even in the most successful cases, the number of transactions involved (low) and the time the scheme has been in operation (short) does not allow any statistically significant findings to be made.

The only firm conclusion, reached in settings as different as Auckland in New Zealand, Montgomery County in Maryland or the commune of Taninges in the French Alps is that prices rise sharply when the procedure is beginning to settle down and then level off or even decline. Could the status of the international property market have had anything to do with this? It is also a fact, however, that acquisition of transferable rights, when not mandatory, places a burden and a brake on the procedure which every effort should be made to prevent.

To be more specific on this important point, there has to be a way to make detailed analyses of local markets in order to set the "market price" of a development right on a countdown basis (from the market price of the end product, the building, less the costs of the operation, deduction is made the highest likely level of land tax and thus the value of the development rights to be purchased). The actual price will probably then be seen to be nowhere near this "market price" unless the purchase of tradable rights is mandatory and there is no alternative nearby (development zone not subject to the transfer system). This comes back to the paradox mentioned earlier, that this system of tradable rights will only work properly in a context where land use is subject to strict planning regulations.

Lastly, the concept of a "real market price for development rights" can be taken even further. Although it can have a meaning at a given point in time, this may prove an illusion when a look is taken at the way the situation may develop. The allotment of development rights is carried out by way of a social and political mechanism that is forced on the players in the market. A public authority may well – as is frequently the case – decide to restrict or even freeze urban development, with foreseeable repercussions on land prices. Unless other corrective measures, which will be described later, are applied, the price of the development right will rise indefinitely. The price of the development right, even though it may reflect the point of equilibrium between supply and demand at any given time is overestimated as a rule by urban development, which is bound to change.

4.6. Equity and efficiency

On the subject under review, the concept of equity has to be considered from the point of view of landowners and from the point of view of the inhabitants of a zone as a whole.

In the first case, tradable rights fulfil an essential function in the absence of a fiscal system capable of removing added value. The price of land is very dependent on the development rights allowed by the zoning regulations, tradable rights making it possible to correct inequities introduced by zoning.

If the concept of equity is extended to all inhabitants, assessment of the method becomes more difficult and depends on the way property rights are conceived and the tax system allowing them to be put into operation. There are two contrasting situations.

In some countries, such as in North America and south-western Europe, there is no universally applicable mechanism for recovering capital gains by urban development and/or payments for development rights. The practice of transferring development rights is equivalent in such cases to distributing the overall capital gain by urban development among landowners only, whereas it might well be expected to return to the community as a whole, in particular when the public amenities that give rise to that added value are funded by the taxpayer. Introducing rights transfers is thus, roughly speaking, a means of formalising a transfer from those acquiring houses to landowners. This is a limited view of equity, which may admittedly be of help in particular cases, but which has no general application in this context. Unless, which is unusual, land is divided up in a very comprehensive and equal way among the inhabitants.

The second type of situation, which is found mainly in northern Europe, is founded on the principle that the capital gain by urban development should return – at least in very large part – to the community. Using various methods (long-term public land holdings in Sweden, municipal monopoly on the supply of building land at controlled prices in the Netherlands), the initial procedure that increases the value of land essentially benefits the community rather than the landowner. The equalisation made possible by transferable rights therefore serves no purpose.

We are thus brought back once more to the definition and content of property law, the key to the problem.

As noted by Donal Krueckeberg (Krueckeberg, 1996): "Property is not just the object of possession or capital in isolation, but a set of relationships between the owner of a thing and everyone else's claim to the same thing. This understanding of property highlights considerations of distributive justice that are particularly important in light of the issues in the contemporary debate about property rights. Rights to personal use of property are fundamental to individual and social well-being: rights to profit from property, in contrast, have always been subject to reasonable constraints for the benefits of the entire community and society. Attempts to establish a contrary case by appealing to natural right, market necessity, liberty, social utility, or just desert all fail to withstand scrutiny. ... These concepts of use rights and profit rights in property are at the heart of the planning question".

In summary, a series of conditions must apply in order to be able to apply a TDR program.

First, the method will only work if there is a demand for development rights, in other words if there is added value to be shared out, and also an urban development plan authorising development.

Moreover, incentives are necessary (bonus zoning) where transfer is not mandatory, and receiver and transmitter zones must be sufficiently large to allow the development rights up for exchange to reach a critical mass. A special body is needed to ensure that the system is properly managed, to keep records of supply and demand, to follow developments over time, to monitor the use made of plots closed to development, and to ensure that price information is transparent. Such a body should help to keep transaction costs down. The public authorities (or a special body) should take on the task of operating a development rights exchange, operating a development rights bank and perhaps regulating rates. Demarcation of conservancy areas (transmitter zones) must be based on clear and explicitly stated criteria in order to gain the active support of the local inhabitants.

The preceding pages reflect a degree of disenchantment with a method that was brought into being with great hopes. Although, under some conditions, the theoretical virtues of bringing the tradable rights system into widespread use seem attractive, more rigorous reasoning and the trials made of the method seem to restrict its scope.

The method has made it possible to solve particular problems. The most appropriate targets would seem to be highly attractive areas where owners of second homes wish to protect their environment without losing the right to sell plots (or rights) at a good price in a context of strong demand and hence of rising land prices. This is the case, for example, of Lake Tahoe or the Praz de Lys plateau at Taninges.

However, widespread use of the method, in addition to a number of technical and institutional difficulties that are hard to overcome, would have distributive effects that could be contested on ethical grounds.

Firstly, the only examples where the system has worked well are those where the tradable rights procedure has been closely tied to a strict land-use plan that has remained relatively unchanged over time. Where zoning is too flexible, the procedure would end by losing credibility.

Another requirement to make the system work is careful and sensitive management. This has generally been achieved in the examples given, but the many unsuccessful cases where this condition has not been fulfilled should not be forgotten.

The (limited) success of the method in the United States is due to the specific legal content of its property law and above all to its case law on added and lost value. The radical differences in practice with West European countries in particular may partially explain the difficulties of transferring the system.

The last word may be left to Ann Louise Strong, an expert in the field, who showed great interest in and enthusiasm for these schemes at the outset. She now has a more reserved outlook: "Although there has been some success with TDR, we are skeptical of the technique in general. It has generated far more discussion than actual land preservation. There are serious practical and legal obstacles to implementing it. ... The author believes that the limited success stories, such as the one in Montgomery County, Maryland, results from regulation, with TDRs as only one small part of the programme" (Strong *et al.*, 1996).

NOTES

1. See Barbieri C.A. (1998); Micelli, E. (1999).

REFERENCES

Babcock R.F. and Siemon C.L. (1985), *The Zoning Game Revisited*, Oelgeschlager, Gunn and Hain, Boston.

Barbieri C.A. (1998), "La perequazione urbanistica: quattro ragioni per la riforma. urbanistica", *Informazioni*, 157.

Bosselman F., Callies D. and Banta J. (1973), *The Taking Issue. a Study of the Constitutional Limits of Governmental Authority to Regulate the Use of Privately-Owned Land Without Paying Compensation to the Owners*, Council on Environmental Quality, Washington D.C.

Callies D. L. (ed.) (1996), *Takings. Land Development Conditions and Regulatory Takings After.*

Dolan and Lucas, American Bar Association, Chicago.

Chung L.W. (1994), "The Economics of Land-Use Zoning: a Literature Review and Analysis of the Work of Coase", *Town Planning Review.*

Coase R. (1960), "The Problem of Social Cost", *Journal of Law and Economics*, 3: 1-44.

Cohen De Lara M. and Dron D. (1997*)*, *Evaluation Economique et Environnement dans les Decisions Publiques*, Rapport au Ministre de l'environnement, la Documentation Francáise, Coll. des Rapports Officiels.

Comby J. and Renard V. (1996), *Les Politiques Foncières*, PUF Coll. "Que Sais-Je", Paris

Costonis J.J. (1973), "Development Rights Transfer: an Exploratory Essay", *Yale Law Journal*, 83: 75-128.

Costonis J.J. (1974), *Space Adrift: Landmark Preservation and the Market Place*, University of Illinois Press, Urbana.

Delache X. and Gastaldo S. (1992), "Les Instruments des Politiques d'environnement", *Economie et Statistique*, Oct-Nov.: 258-259.

Falque M. and Massenet M. (eds.) (1997), *Droit de Propriètè et Environnement*, Dalloz, Paris.

Fischel W.A. (1978), "A Property Rights Approach to Municipal Zoning", *Land Economics*, 54, (1).

Hagman D. and Misczynski, D. (1978), *Windfalls for Wipeouts, Land Value Capture and Compensation*, Aspo, Chicago.

Henry C. (1976), *Mècanismes de Solidaritè Foncière et Amènagement de l'espace*, Cahiers du Laboratoire d' Econometrie de l'Ecole Polytechnique, Paris, Mars.

Johnston R.A. and Madison M. E. (1997), "From Landmarks to Landscapes, a Review of Current Practices in the Transfer of Development Rights", *Journal of the American Planning Association*, 63, (3).

Krueckeberg D.A. (1995), "The Difficult Character of Property. To Whom Do Things Belong?", *Journal of the American Planning Association*, 61, (3).

Lenotre-Villecoin J. (1975), "Urbanisme et propriètè du sol. Les propriètaires vont-ils pouvoir monnayer des droits de construire fictifs?", *Les Etudes*, Avril.

Micelli E. (1999) "Les Droits Negociables dans la Gestion des Nouveaux Plan d'urbanisme", *Etudes Foncières*, 82, Mars.

Moore T. D. (1998), *Transfer of Development Rights in the United States. The Experience in two Selected Jurisdictions*, In Press Estudios Publicos, Centro de Estudios Publicos, Santiago, Chile.

Pizor P. J. (1986), "Making Tdr Work, a Study of Program Implementation", *Journal of The American Planning Association*, 53, (12).

Pruetz R. (1997), *Saved by Development: Preserving Environmental Areas, Farmland And Historic Landmarks with Transfer of Development Rights*, Arje Press, Burbank, Ca.

Renard V. (1980), *Plans d'urbanisme et Justice Foncière*, PUF, Coll. "Espace et Libertè", Paris.

Strong A.L., Mandelker D.R. and Kelly E.D. (1996), "Property Rights and Takings", *Journal of the American Planning Association*, 62, (1).

Strong A.L. (1986), "Transfert de Cos Aux Etats-Unis. Le Parc des Pinelands", *Etudes Foncières*, 30, March.

9. TOURISM DEVELOPMENT THROUGH THE QUALITY AREA PLANS

MAGDA ANTONIOLI CORIGLIANO

1. ENVIRONMENT AND COMPETITIVENESS IN TOURISM

In the global competition system the environmental variable is acquiring more and more significance both from the micro–economic and the macro-economic points of view for three main reasons:

- it allows to improve the product at corporate or at area-system level in a growing – standardised environment (thus inducing the consumer to accept an increased price for additional distinctive elements);
- it calls for greater attention and greater "sensitivity" to multiple actors such as firms, public authorities, policy makers and various associations, thus contributing on one hand to step up attention and the instruments at protection system level and on the other hand to create ethical values;
- it helps to overcome the antithetical position of public policies and corporate criteria, especially where the latter gain in competitiveness by eluding regulations.

All above-mentioned factors call for an active management of the environmental variable which is, to all intents and purposes, part of strategic choices and may certainly change the relationships between the single operators and the public sector; it is however necessary to enhance law observance and to valorise and impose coherent and rational restrictions in order to ensure environment preservation. The same is particularly true for the tourist sector, where the environmental variable may not be exclusively dealt with by reducing all negative externals and by highlighting the related social costs. The environmental variable is a factor directly affecting tourist production. It is important to point out that the process final output, that is the *tourist product*, considered in the broad sense of *area product*:

- results from the *co-production* of multiple actors, not only of economic actors;
- features many *immaterial components*;
- is perceived – and assessed – by the *single* tourist *comprehensively*.

The Public Administration is thus replacing its restriction policies – merely aimed at controlling externals – with a more global strategy focused on the support and the promotion of environment innovation. Such action should be backed up by the private entrepreneurial sector which should develop management functions based not only on environmental accounting and on more compatible processes and products but also on an open cultural pattern ready to seize the new concrete opportunities arising.

2. THE ROLE OF LOCAL TERRITORIAL PLANNING IN TOURISM AND IN THE ENVIRONMENT

The environment issue has a global dimension by definition, and is widely affected by the ecological balance of the planet. But it is locally that intervention policies and operative instruments are to be developed at different territorial levels. It is not possible to define univocally the conditions for a sustainable development that cannot be ensured by the sum of partial balances. Subsequently there are no measures applicable to every level of the different ecosystems.

In such process the role of the public actor is paramount and highly dynamic. The basic paradigm long adopted by public decision-makers is undergoing major transformations; such paradigm was based on strong interventionism used to lead the development process and on greater territorial transformations, as well as on a top-down hierarchical planning pattern. Such approach turned out to have a number of methodological and operative limitations in its actual application which determined exasperated conflicts between local authorities, regional bodies and – last but not least – stakeholders both at horizontal and vertical level.

For such reasons, there is a tendency towards strong deregulation, where spontaneous processes are gaining more and more importance and are managed and controlled according to a more negotiating and co-operative concept, thus allowing to involve more private operators even at applicable level. Such process has been significantly facilitated by the growing operational autonomy of the single public actors involved in it and by an approach which proved to be less "sectarian"; the management environment, which was formerly chiefly regional and based on formalised plans, has become more informal and is now focusing on joint actions between Town Councils, different Chambers, companies and private individuals. Such transformation allowed the construction of major transport infrastructures and public works which play a paramount role as far as strategic investments are concerned in a wider process of internationalisation and globalisation.

The tourism sector is obviously more vulnerable to the effects of such planning choices and of environment preservation in general for the reasons expressed in the previous paragraph. The tourism sector suffered both from devastating actions of totally deserted areas and from equally harmful choices based on a too restrictive and authoritative approach.

In brief, the environment considered as a resource and national heritage is to be preserved and enhanced through the qualification of local systems. It is at such level that naturalistic and economic interrelations and interdependencies may be highlighted within a wider context of evolutive and systemic preservation, positioned at a higher institutional level, where goals are established and an identity of direction and regulation is guaranteed.

Adopting such co-operative and negotiation approach in territorial planning in harmonious balance with tourist development guidelines / trends calls for a joint effort combining scheduling and productive aspects, involving all actors – considering how the decision–making system is fragmented and widespread – and focusing on quality. Quality intended not merely applied to services or to other partial components of the "tourist product" but as quality of the tourist area.

In the next paragraphs the main concepts underlying the definition of area quality of a tourist resort will be pointed out together with their subsequent policy implications.

3. THE TOURIST DISTRICT AS THE DIMENSION FOR THE AREA QUALITY

3.1. The concept of "tourist district" or of "territorial system"

The issue of quality applied to the tourist product in terms of – product-area – should take into account not only the role of territorial resources, but also the extra-economic relationships existing between different groups of operators: between operators and stakeholders, as between firms and the environment. It is thus advisable to adopt the district as point of reference in order to better assess the negative or positive impact of extra-economic relational dynamics on quality. The concept of "district" is aimed at identifying all areas where firms dealing with tourist activities have achieved internal organisational skills and such a degree of development as to generate competitive advantages within the relational networks existing in the considered area.

There are multiple implications arising which deserve further examination considering that:

- internal economies and some external economies for the single firm become internal economies for the network;
- only the external relationships of the network are analysed, the single firm's are not considered;
- the most significant competitive advantages are related to the network as a whole and not to the single firm.

3.2. Definition of district

Broad literature is available on districts but is mainly focused on the activities of industrial districts, while the purpose of our discussion is to identify only those aspects which are useful to define the tourist district. We share Beccattini's approach (1979) who defines the district unit as "intensifying localised relationships between firms reasonably steady in time". In such definition, the connection between firms and the territory is essential; subsequently the territory can be also considered as a "socio-territorial entity – a limited area with naturalistic and historical heritage – featuring an active coexistence between a community of people and a population of firms". The coexistence of utterly different elements – history, culture, the different profiles of the firms involved, the different interactive networks established – makes each district unique; as a matter of fact tourist areas in general differ significantly from each other from the cultural, historical, productive, qualitative and environmental point of view. Nevertheless, the differences existing within the various district systems do not seem to be relevant, but might be considered as "variations" on a common track whose cardinal factors are the following:

- each *determined* and *limited* geographic area, covering the territory of one or more municipalities, often has its own specificity as far as the conformation of the territory and the ethnic origin of the population are concerned;
- the presence of *specific* and *characteristic* elements of historical, cultural and social identity and of economic-productive patterns, typical of the local community which fosters the creation and the transmission of specific tourist culture;
- the role of *local institutions* providing services and appropriate structures and infrastructures, promoting initiatives aimed at managing plans for entrepreneurial development; such role indicates the high level of involvement of all social actors to guarantee the liveliness of the district itself;
- a *population of firms* connected interactively and interested in implementing a formalised network of relationships, with different hierarchical degrees.

In the district firms are *deeply entrenched* in the area, in the history, the culture and in the social patterns of the local community. Such characteristics create a strong sense of belonging to the local system in the operators; thus the co-ordination of activities is simplified and inter-firm and interpersonal relationships are enhanced. In such a context, local transitions – traditionally based on market economy – are also affected by extra-economic factors of the area and the environment, which includes such components comprehensively, becomes a fundamental variable affecting deeply local productive processes.

Another important feature characterising district firms is the formalisation of relationships and thus the determination of specific roles. Formalised relationships imply the collocation of each firm at a specific stage of the productive process typical of the district considered, thus creating a *complementary relationship*. Such connection – which generally produces an actual "organic interdependency" reinforces the sense of belonging in terms of role and collective responsibility, each actor being aware of the economic benefits arising from it.

3.3. District as an economic pipeline

Within the analysis of districts, the configuration of economic pipelines – as far as the production of value and the chains of value are concerned – is particularly important. The district can also be seen as a territorial agglomeration of specialised and integrated firms situated within specific productive pipelines.

The analysis of each pipeline can be integrated with the reticular analysis of the district which enables to highlight the multiple upstream, downstream and horizontal relationships between firms. Such instrument is essential in the current context in particular where the competitiveness of a firm does not depend only on its resources but also on the external conditions of the entire socio-economic system where it is operating.

In industrial firms the *upstream stage* consists in selecting and supplying raw materials. For tourism "raw materials" are represented by supplies for receiving pipelines, catering facilities but also tourist resources and relevant activities such as snow activities and ski facilities. One peculiarity of the district is its attempt to progressively internalise upstream specialisations, the real asset of district economy. It is at this stage that the most important scale, specialisation and agglomeration economies are created.

Downstream stages consist in market analysis, promotion, distribution and post-sale assistance. At such stages small firms prove to be weak: their control generally requires much wider scale dimensions than small district firms are able to offer; furthermore the process of intermediation is often dominated by monopsonistic structures. Despite such difficulties, downstream stages arouse major attention since an efficient expansion policy represents the essential condition for the global development of the system. For such reason the direct or associate control of downstream stages by district firms is acquiring more and more importance.

The role of such downstream openings explains the progressive internalisation of upstream specialisations: such phenomenon is possible only thanks to the presence of a global concept combining the downstream cycles. In fact, the internalisation of upstream stages enables the development of new competence and allows the district to increase its competitiveness, thus creating a virtuous cycle.

3.4. The tourist district

The main features of district areas hitherto presented are essential for the tourism sector especially as far as the connection with the territory, the specialisation of productive processes and the development of pipelines are concerned. In tourist production, one of the most crucial and elusive factor necessary to identify an area as a district – that is defining a specific culture and a local peculiarity – turns out to be significantly simplified since the attractiveness of its resources – which are so indissolubly bound to the territory and to its cultural heritage – already identifies a potential district area. The concept of "district" though should be corroborated by other parameters such as the productive dimension, the number of firms, the number of formalised relationships, the role of local institutions. We do not intend here to provide an alphanumeric classification of all variables relevant to assess the district features of an area, but we would like to point out how the tourist district differs from an industrial district:

- the role of *background networks* (extra-economic networks) is significantly more marked and is determined by the pre-eminent role of the local community and of its institutions in the tourist product, as well as by the complex issue of natural and cultural resources to be included in tourist supply in terms both of productive processes and of preservation;
- the *horizontal structure* of the economic network between the operators of the same component of the tourist product, featuring a growing degree of competitiveness (antithetical networks) instead of synergy (which has a significant bearing on small firms operating in the tourist sector);
- a complex interlacing competence network (*decision-makers' network*) at different levels of the territorial hierarchy even aimed at financing promotion and investment programs;
- the *extraterritoriality* of many actors (decision-makers) of the tourist production process (of tour operating in particular).
- The most critical point is certainly the link between the different structures of tourist supply and, in particular, between public and private operators. Such relationship is essential to consider the issue of the

management of resources within the tourist product that are to be preserved and used in order to create the tourist product. Furthermore, such relationship is responsible for the creation of almost all "district economies" such as:

- scale economies in promotion and commercialisation at corporate, local or national level;
- specialisation and scope economies in production and in connection with the markets (i.e. relationships with intermediation bodies, price policies for direct sale channels, analysis of information within market research);
- agglomeration economies.

Such crucial points call for the development of appropriate forms of co-ordination, which are paramount considering the peculiarities of the tourist sector: the simultaneity of production and consumption stages – typical of horizontal pipeline productive processes, the impossibility to detain stocks, the strong involvement of human resources in production and supplying stages, last but not least the vulnerability of the tourist sector to economic conjunctures due to its globalisation and to the horizontal character of the economic networks. Co-ordination is the key in tourist areas as far as area quality is concerned since it represents the most efficient solution – and often the *only possible solution* – able to combine the different orders of priorities adopted by tourist operators when short-term strategies are preferred to medium and long-term planning – which represent the necessary time span to carry out development analysis – behaviours focused on personal activities are adopted without considering the possible consequences on the local tourist system. It is crystal clear that only a system-oriented organization able to trigger off short, medium and long–term action strategies may be able to plan and implement an Area Quality Plan for one or more (homogenous) tourist resorts which often present a plurisegmentation of the product at typological level (seaside and mountain resorts, congressual, thermal and cultural sites) or at seasonal level (summer, winter).

4. AREA QUALITY IN TOURIST DISTRICTS

Unlike the approach adopted for the district theory, it is very difficult to apply the same development methodology of implemented quality in the industrial sector to the service sector in general and to the tourism sector in particular. Such methodology may be applied, with a number of appropriate changes, only to those structures offering tourist services that are somewhat similar to industrial firms for their internal production cycle such as hotels (hotel chains in particular), holiday villages, tour operators, information offices. Such structures have often adopted quality certifications – often in compliance with ISO 9000 regulations – or codes of practice aimed at ensuring certain guarantees to the customer. In such cases, as for all firms offering services, quality is referred to the *process* and not to the *final product*.

However, if referred to the entire area, in the broader sense of global tourist product, the quality concept acquires a completely different connotation that loses its corporate feature and inevitably involves the behaviour of the tourist, of other tourists, stakeholders, public and private operators comprehensively. On such premise, defining a quality process at tourist destination level implies remarkable conceptual and operative difficulties. In any case, the certification methodology does not seem to be eligible in that identifying the processes an area quality should adopt and the regulations it should comply to is very difficult. On the contrary, quality planning results to be essential and it consists in defining the elements supply and resources should have to guarantee the tourist a minimum service standard. In other words, certification refers to the single component of the tourist product (such as hotels, restaurants) while planning refers to the area comprehensively.

5. QUALITY AT TOURIST DESTINATION LEVEL

Giving a full definition of the term "quality" referred to a product-area means having an answer to the following question: is it possible to guarantee the quality of a tourist destination?

Nowadays offering services to the customers meeting their expectations does not necessarily mean to have success in tourism. It is assumed that if such aspect lacks, it is a sign of non-quality; but if it is present, it does not automatically guarantee quality either. Quality means that the customer is globally satisfied with the services, eager to come back, and to recommend the product to others.

Consequently, our discussion on quality cannot be focused only on instruments, infrastructures and services but also on the *tourist environment* and on all its components such as the population and the naturalistic aspects. In fact in the quality of the tourist sector goes beyond the difference between the services expected and the services received but it is the result of an experience process undergone by the consumer leading to global trust in the area-destination ("experience and credence quality").

It is however important to underline that despite quality is an integral part of a tourist product, it is also affected by the *cost/service* ratio. But to what extent is the customer willing to pay an additional price for additional quality ("willingness to pay")? According to the Anglo–Saxon experience, in case of comparison between supply and its price the customer tends generally to choose the *most favourable* offer which is not necessarily the *most economical* according to the "value for money" principle. *Value = price* × *quantity* × *quality*. Such statement is supported by various tendencies expressed by most of international tourist demand, whose main features are: elasticity to return and inelasticity to the price. Thus, the segmentation of demand – and consequently of supply – on the different price levels requires the adoption of *different forms of quality* more or less dependent on the price: each level has its quality. Quality may exist at any price level since the customer that chooses one product instead of another expects a different level of quality.

Advantages of the Area Quality

In general, adopting a quality strategy at tourist district level produces the following advantages:

- as fully demonstrated by the industrial sector, the quality process obliges to define the goals of the economic action to be carried out (tourist model), to design a plan to achieve them, to constantly control its application avoiding any dispersion (waste) of energies, time and financial resources, in other word it requires joint actions;
- it is easier to highlight non-quality costs such as necessary costs (which are not few) in order to acquire new customers especially when the fidelisation of old customers was not successful;
- it represents an important marketing and promotion instrument for the image;
- it contributes to improve competitiveness factors and the position on the

market of the single actors thus increasing investment profitability;

- it creates significant benefits to the residents too (it improves the quality of life standards);
- it guarantees development and positive benefits to all the economic activities of the area.

The ultimate goal of quality for a tourist destination is not only to satisfy its customers but also to improve the quality of life of its residents and to develop the economic activities of the area.

Perception of the area quality

The perception of quality by tourists/consumers develops along three levels:

- the *compulsory function* which refers to base quality. Its lack is inadmissible, its presence is not sufficient to talk about quality of the tourist product;
- the *progressive function* (or use function) where quality perception is directly linked to the level of services received;
- the *attracting function* which offers an unexpected but gratifying service, a plus product turning the customer's experience in a dream to remember and to tell others.

The main factors determining the users' opinion are: comparison with competitors; personal tourist experience; commercial promises as expressed in communication and promotion messages.

However, operators do not always find it easy to understand demand's perceptions as at the tourist destination level different perceptions superpose and interlace themselves. In particular accepted non-quality, non-accepted non-quality, convincing over-quality and useless over-quality are not easily distinguishable.

Main critical points connected to the process

In general the main difficulties connected to the adoption of a Quality Plan at destination level are the following:

- it requires the co-operation between many firms and organisations very

different from one another;
- public and private organisations, associations;
- large, medium, small enterprises;
- experienced and inexperienced operators;
- qualified and non-qualified personal behaviours;
- seasonal and permanent operators;
- a common quality criterion;
- common goals;
- the implementation of a joint-action programme applied sectorially;
- the acceptance of certain common benefits, but unequally distributed .

This last aspect calls for specific patterns to manage the non-homogeneity of economic returns; at tourist destination level a quality process can determine neither the same level of economic return for each operator nor a return proportional to the improvement efforts of everyone. In particular the main critical points are the following:

- avoid that the single operators of the area are tempted not to subscribe a common agreement because they are convinced to profit from others' intervention without bearing any cost;
- appointing a quality controller;
- appointing the sanctioning authority for punishing single operators' behaviour which do not comply to collective decisions and ruin the global image of the tourist destination and harm other operators' efforts.

European experiences with the implementation of quality plans, in France in particular, showed that the co-operation between different actors is the real critical point; apart from being a demanding effort in terms of time, energy and money, it often upsets habits, privileged positions and personal advantages. For such reasons, an accurate definition and implementation of Quality Plans are necessary taking into consideration the following aspects.

A quality plan is accepted with conviction especially when the tourist destination is obliged to. The resorts mostly convinced of quality plans are non-dominant stations since they are less solid in terms of image, less competitive on the international market, more vulnerable in case of economic recession or bad weather conditions. On the contrary important resorts are not very willing to revise their strategies.

It is easier to implement an endogenous quality process rather than importing a new method borrowed from other experiences. It is difficult to get economic operators to accept the price of quality. Public financing contributes to introduce it; however local operators perceive more immediate benefits from funds allocated for communication and promotion initiatives without considering:

- non-quality costs which mean ever-growing costs for communication and promotion activities necessary to acquire new customers;
- the economic value added determined by quality (fidelisation, free-of-charge communication thanks to the pass-the-word concept);
- pipelines are more sensitive than stations·

Within the pipeline, there is greater awareness that the process quality can be harmed by the non-quality of one single ring of the chain; consequently there is greater willingness to impose quality criteria (possibly, even certification) within the formalised relationships arising within the pipeline itself.

The same quality standard can rarely be guaranteed throughout the year.

Quality standards evidently refer to a certain value (or range) of numerousness of consumers; such condition is not satisfied due to the extremely variable tourist flows over the year. Besides, peak values of tourist flows cannot be taken as reference values since they would automatically engender an over-quality situation which would not be economically acceptable and would not produce corresponding and proportional benefits to the tourist resort.

6. THE ROLE OF INTERMEDIATION IN THE AREA QUALITY

In this paragraph the aspects connected to intermediation will be discussed with particular reference to the role travel agents and touring operators play in defining the quality of a tourist destination. On one hand they are addressing demand which is more and more environmentally-conscious, and on the other hand they are open to a wider system of offer where they may transfer and impose their tourist services (cultural, eno-gastronomic, artistic, sport activities and natural resources). Respecting the environment and residents' needs – who are their first allies in a product-destination concept oriented to quality – helps to reinforce the central position of such actors on the market.

In general, intermediation activities based on area quality shall support their actions through initiatives concerning sustainable development such as:

- creating/selling of highly-qualitative packets with strong environmental impact as defined by the Plan; and eliminating all packets with strong negative impact such as hunting endangered species, tourist circuits in particularly vulnerable areas;
- supporting and participating in initiatives aimed at preserving the environment and eco-sponsoring;
- adding environment-connected material to the ordinary brochures;
- choosing environmentally-conscious suppliers (hotels, ground operators) and requiring targeted environmental standards;
- including competence and assigning particular attention to environmental issues in the corporate image;
- urging local authorities of destination areas to adopt appropriate measures in order to achieve more significant qualitative standards[1];
- introducing training activities for employees and for the agent/distributor network to foster and enhance environment-consciousness.

7. CONCLUSIONS

Our discussion started by analysing environmental planning within the tourist factor. The tourist factor represents the basic conceptual framework necessary to define a tourist product considered as product-area and provides the more appropriate approach to examine environmental aspects in tourism such as environment preservation – and consequently sustainable development – and economic effects on the activity of tourist operators in the short, medium and long-term. An operative approach has been chosen in order to lay out efficient proposals that policy-makers can immediately put into practice at local level.

Our analysis has developed along two main guidelines:

- highlighting the indissoluble bond between the environment and the concept of quality in the tourist area;
- selecting an analytical methodology for districts of tourist areas, since it

represents the most efficient instrument to highlight the critical points arising from the quality management of a tourist product at area level.

The first point required a definition of area quality and of its structural features. Unlike the type of quality typical of industrial productive processes, area quality cannot be assessed only in terms of gap between expectations and services received by the consumers. The quality of the relationship between tourists and their destination is not exclusively determined by the services offered, but it is an actual experience in which public and private operators, the local community (stakeholders) and the tourist themselves are involved. Taking into account the experiences of different countries, we have concluded that the management of quality in a tourist area shall not develop itself through certifications – whose efficiency is excellent on a single service (such as receiving and catering facilities) in terms of costs-benefits while proved to be weak if applied to the whole area – but through Quality Plans. Quality Plans should outline the optimal features necessary to guarantee that the tourist product as a whole efficiently satisfies the demand's needs.

The management of Quality Plans is not at all easy, because it requires the co-operation of actors with completely different and often contrasting orders of priority. For such reasons, the creation of a plan requires two kinds of actions:

- the implementation of a co-ordination instrument (Service Centre) above the parts able to combine the different orders of priority in particular private operators' short-period strategies with public operators' long-term strategies but also to translate into action all potential scale and system economies of tourist district areas;
- bring area quality – and its environmental aspects in particular – to the attention of external operators, that is intermediation by tourists themselves. Such point is strictly connected to the previous point: the existence of a co-ordination centre situated within the tourist area itself is able to bring attention to the area more efficiently.

The last point highlights the importance of tour managers whose role is still rarely present on the market. Agencies and tour operators play a paramount role in defining and controlling area quality. The eco prefix does not prove to be successful in the long term. A real *ecological value of the product* is necessary and in this sense the travel agent, together with the tour operator, can act on the tourists' behaviour, can communicate environmental value through material and information and can lead targeted communication campaigns on environmental initiatives. These are all factors indicating an active participation to the quality process, such action will be even more efficient if included in an actual Quality Plan. In order to realise such competitive advantages, the adoption of the aforesaid measures shall be combined with the subscription of codes of practice – also known as service charter – that would allow demand and operators to recognize their own strategic orientation towards quality. In some cases, especially when some components within the destination adopted the quality certification – whether an Area Quality Plan is present or not – it can be advisable for the intermediator to acquire, despite its costs, the certification instead of the codes of practice since it can become a requirement service suppliers themselves impose to the distributor. In such process the agent's closest allies are the managers of the packet's different components, that is those who participate to the Area Quality chain in destination tourist resorts. De facto, at operative level, such terminals support in the first person, even as far responsibility attribution is concerned, the agent for product qualification and consequently for the protection of the user/tourist.

NOTES

1. Such case is emblematic: one tour operator is threatening to rescind all agreements drawn up with Venetian receiving firms because they lack of water-treatment facilities.

REFERENCES

Antonioli Corigliano M. (1988), "Uno studio microeconomico sulla domanda turistica e delle implicazioni di metodo per la politica del turismo", *Commercio*, 29.

Antonioli Corigliano M. (1989), "La tematica ambientale nella pianificazione turistica a livello locale: un tentativo di razionalizzazione", in *Turismo e ambiente nella società postindustriale*, FAST – Touring Club Italiano, Milano.

Antonioli Corigliano M. (1990), "L'analisi di impatto ambientale nelle decisioni di investimento", in *Elementi per un'analisi degli effetti economici di un attraversamento stabile dello Stretto di Messina*, Edizioni Scientifiche Italiane, Napoli.

Antonioli Corigliano M. (1994), "Tourism and sustainable development: theoretical issues, impact assessement and environmental policies", *Turistica*, January – March.

Antonioli Corigliano M. (1994), *Il ruolo di raccordo tra attori pubblici e privati dei Centri di Servizi al turismo, con riguardo alle attività di investimento e promozione, alla luce delle iniziative comunitarie*, CNR – Progetto Strategico Turismo, mimeo.

Becattini G. (1979), "Dal settore industriale al distretto industriale, alcune considerazioni sull'unità di indagine dell'economia industriale", *L'industria. Rivista di economia e politica industriale*, 1.

Becattini G. (ed.) (1989), *Modelli locali di sviluppo*, Il Mulino, Bologna.

Brent Ritchie J. R. and Crouch G. I. (1997), *Quality, Price ant the Tourism Experience: roles and contribution to the destination competitiveness*, Proceedings of the AIEST annual meeting "Quality and Tourism", St. Gallen.

Campari I., Mogorovich P., Montanari A. (1989), *Cultura dell'informazione e gestione dell'ambiente*, Edizioni Scientifiche Italiane, Napoli.

Casarin F. (1995), "I rapporti di collaborazione tra impresa alberghiera e impresa tour operator", *Economia e diritto del terziario*, 3.

CERTeT – Centro di Economia Regionale, dei Trasporti e del Turismo, *Cooperare per competere: le politiche regionali per la competitività dei distretti industriali*, Workshop 9 June 1997, Università di Bocconi, Milano.

Consulting and Audit Canada (1995), *What Tourism Managers need to know: a practical giude to the development and use of indicators of sustainable tourism*, WTO, Madrid.

Cossentino F., Pyke F. and Segenberger W. (1996), *Local and regional response to global pressure: The case of Italy and its industrial districts*, Research Series 103, ILO publications, Geneve.

Depperu D. (1996), *L'internazionalizzazione delle piccole e medie imprese attraverso aggregazioni. I consorzi export*, Ricerca di base, Working Paper n. 7, april, Università Bocconi, Milano.

Goodman E. and Bamford J. (1989), *Small firms and industrial districts in Italy*, Routledge, London.

Keller P. (1997), *La gestion de la qualité dans le domaine du tourisme: question à traiter*, Proceedings of the AIEST annual meeting "Quality and Tourism", St. Gallen

Handszuh H. (1997), *Communicating Quality to Tourism Consumers*, Paper submitted at the International Conference "Competing in Tourism Through Quality," Venice, 12–13 December 1997.

Onida F., Viesti G., Falzoni A.M. (eds.) (1992), *I distretti industriali: crisi o evoluzione?*, EGEA, Milano.

Origet du Cluzeau C. (1997), *La qualité a l'echelon de la destination touristique*, Proceedings of the AIEST annual meeting "Quality and Tourism", St. Gallen.

Piore M. and Sabel C. (1984), *The second industrial divide*, Basic Books, New York.

Pyke F., Becattini G. and Segenberger W. (1990), *Industrial districts and inter–firm co–*

operation in Italy, ILO publications, Geneve.

Rispoli M. (1995), "L'individuazione dell'ambiente competitivo per le imprese alberghiere", *Economia e diritto del terziario,* 3.

Senn L. (1991), "Servizi strategici per le imprese". in *Impresa e Stato, Rivista della Camera di Commercio di Milano,* 14, June.

Senn L. (1998), *Ambiente e Pianificazione Territoriale,* Proceedings of the Conference "Ambiente e competitività", Università Bocconi, Milano 20 febbraio 1998.

Sereno A. (1995), "Evoluzione dell'intervento pubblico nel turismo", *Economia e diritto del terziario,* 3.

Thorelli H.B. (1986), "Networks: Between Markets and Hierarchies", *Strategic Management Journal,* 7.

Troilo G. (1998), *Il green marketing: il ruolo della comunicazione ecologica,* Proceedings of the conference "Ambiente e competitività", 20 February 1998, Università Bocconi, Milano.

Vaccà S. (1998), *Istituzioni e strumenti nel governo dell'ambiente,* Proceedings of the conference "Ambiente e competitività", 20 February 1998, Università Bocconi, Milano.

Vanhove N. (1987), *Tourism Quality Plan: an effective tourism policy tool at the destination level,* Paper submitted at the International Conference "Competing in Tourism Through Quality," Venice, 12–13 December 1997.

10. SOCIAL EXPECTATIONS AND SUSTAINABLE TOURISM DEVELOPMENT: THE TERRITORIAL PACT

FRANCESCO LOSURDO

1. INTRODUCTION

There is a general consensus that the industrial era, characterised as it is by the continuous growth of mass production (i.e., goods destined to direct consumption and intermediate goods instrumental to the production of other goods) and by the uncontrolled large-scale consumption of resources (soil, energy and water) is close to its end. In fact, the present stage of economic development is already referred to as the post-industrial era, an era where production should mostly be based on the use of immaterial goods (services, information, technologies) as well as on the re-utilisation of remains of production and consumption such as secondary raw materials.

At least in the developed countries these means of production and consumption are by now widespread and shared between producers and consumers, in addition to being imposed by the pressure of competition and by the growing awareness of the limits of environmental resources.

The issue of opting for either the industrial or the post-industrial growth paradigm is, instead, present in the developing countries – which should perhaps skip over an entire phase of economic history – and, in a different manner, in the disadvantaged regions of the industrial countries. These regions are, in fact, asked to make a difficult choice between completing the industrial phase of development – that seems, at least in part, to have already reached its maximum expression – and diverting the enormous financial flows destined for regional development policies to prioritise the restructuring processes meant to turn to a post-industrial economy where education and training should be focused on lighter production techniques, information, research and development activities, cultural and training services, proximity services, services for the enjoyment of natural and art resources, and tourist services.

This evolution of production and consumption models, together with the issue of having to choose between two alternative development models in different areas for the degree of economic development, the different structures of production and or organisation, and the endowment of factors and resources give a less abstract meaning to the perspective of sustainable development, considered by now a world-scale target; and at the same time they tend to place in a local context the relationships between the economy and the environment, and this not only disseminates knowledge and awareness among the economic actors and society as a whole, but also ensures that, after having assessed the existence and extent of the environmental damage, and promulgated regulations on a world scale, the task of selecting the development model and organising rule of sustainability is mandated to the regional governments.

This apparently clear-cut sharing of tasks is countered by very complex choices and inefficient operational models adopted to assess the feasibility of the development policies in environmental terms. As a result there is often dissatisfaction of an ideological nature which, sometimes also instrumentally, determines stalemates in the process of decision-making and the working out and implementation of development programs.

Political decision-makers – generally without a real understanding of the environmental problems concerning economic growth – believe that underrating this issue might mean a loss of consensus on the part of their knowledgeable voters; public officials fear breaching rules and laws – which they sometimes are not familiar with because of the great proliferation of rules in this field – thus fearing to trigger uncontrolled reactions by environmentalist groups and/or negative judgements by courts. The entrepreneurs and the workers are bothered by both politicians and public officials as, by delaying decision-making, they are thought to harm production and employment.

In short, the issue of the environmental sustainability of the development policies and programs remains a factor of social disaggregation and a cause of overt or hidden conflict amongst the various groups in society. This paper is based on the concept that the prerequisite for opting for sustainable development is the choice of a post-industrial type of economic development which can serve as a first step in the establishment of a new form of *limitless development,* based as it will be on information and intelligence, entities that, by their very nature, are limitless[1], and on lean production. In addition, we try to demonstrate that the objectives of sustainable development, in particular in the tourist sector, must be placed in a regional context in order to be implemented as efficient operational programs, not only because environmental resources are endogenous to the region where tourism should be enhanced, thus differing from region to region, but also because the decision-making and the operational program elaboration skills, the potentialities of implementation of the programs, the avenues and modalities of growth, the relationships among the bearers of ideas of needs and interests are highly differentiated at the different regional scales. Therefore in order to be feasible, a sustainability model has to be *embedded* in the regional dimension and based on the self-interpretation organisation skills of the individual territorial systems. Given the still blurred contours of the concept of sustainable development, we have decided to define sustainability lying between weak sustainability[2] and strong sustainability starting a consideration that a society's development level, assessed in terms of social well-being – in other words degree of satisfaction of needs - is composed of flows of utility made by men, through the organization of factors or relationships, and of flows of utility offered by nature.

The problem consists in determining the threshold of well-being below which there is no development, in the sense that the total social well-being would diminish because natural flows of utility do not compensate for a decrease in the product flows; and the threshold above which there is no sustainability, in that the increase in well-being would be obtained thanks to the fact that the product flows with utility absorb part of the natural utility flows without compensating them. The feasibility of strategies of this kind impose a formative-informative type of approach that tends to reduce the informational asymmetries and support the aggregation of interests disposed to pay in terms of renouncing present well-being (produced component) in exchange for a greater future well-being (natural component). A strategy of this kind, albeit only sectorial, should possess:

- a well-defined regional basis, as a context for decision-making in terms of models of development, of the location of environmental resources, of the location of the industrial, exchange and relationship processes that are established with the resources of the environment;
- a method to understand interests and to monitor those human behaviours that are in a direct relationship/opposition with the environmental system;
- an implementation model that relies on criteria and parameters which can suggest, assess and monitor the *externalities* of the production

systems.

As these variables are articulated and differentiated on a local scale, the local system seems to be the most adequate dimension to work out efficient operational models for a sustainable development.

Of course this operational scale is not exempted from a series of risks: the risk of fragmenting the policies of sustainable development into programs which are not sufficiently integrated, especially when the strategy is of a sectorial nature; the risk of suffocating the interest of the weakest; the risk of bureaucratising the use of the resources; the risk of excessive recourse to the veto power or, quite the contrary, the risk of too much permissiveness. All these potential risks are due to the different types of relations and to the multiple interests present in the local context.

To avoid these risks an operational model must necessarily go through a *planning* phase as an approach to the objectives of sustainable development, even of a sectorial nature, with a particular emphasis on the tourist sector.

The *territorial pact* can be the most efficient expression of the social partnership that underlies the planning process. Here we examine the conditions under which the territorial pact could become a feasible model for a perspective of sustainable tourist development, acting, as mentioned, on a local territorial scale.

2. SUSTAINABLE TOURISM DEVELOPMENT IN A POST-INDUSTRIAL ECONOMY

The first theoretical studies on the post-industrial economy move from a radical position, where post-industrial is understood literally as the deindustrialisation and abolition of industrial growth as we actually know it, to a less extreme concept, where a prevalent role is played by labour-saving and environment-saving technologies, and the greater time available is used to fulfil cooperation and voluntary service activities to sustain an appropriate environmental development; and finally to a moderate interpretation that, contrary to non-market and productive autarky hypotheses, contemplates a radical decentralisation of roles to local communities in a context of self-support/self-organisation of resources against needs, to be realised through normal market mechanisms and through currency use[3].

These differences do not prevent us from seeing the clear similarities between the first theoretic expressions of the post-industrial economy and the topics of weak or strong sustainability, of the territorial-scale of growth, of flexible production and of the cooperation/coordination/competition relationship in the productive system of small and medium enterprises. In fact they have the same root in the search for a development model that goes beyond Ford's model based on mass production, with its fulfilled strategies, governmental formalities, and the resultant transformation processes of the economy and society. Nevertheless the same divergences complicate the interpretative scene, so that it seems simpler to refer to substantial expressions of a post-industrial development process and to the more current way of grasping this stage of development.

Generally, the post-industrial stage of development is characterised by a change in the material contents of production and of wealth, to which corresponds a transformation in the organization of work and the social structure. This partial definition often is replaced by *service society* (or *information society*), which, even if more complete, is nevertheless insufficient because it does not consider the necessary institutional and normative adjustments that regulate the production and transaction relationships, the property rights, or conceptual categories not traceable back to the service society idea, such as that of *complex utilities*, to whose formation the environmental use values also contribute.

Given that the interpretative schemes proposed all have a certain summariness about them, we shall try to discuss the actual transformations in the economy, in the environment and in the society linked to post-industrial development, from which we can deduce some solid and useful indications for viewing sustainable development. These are:

- a progressive reduction of the employment of manpower by *machine-intensive* productions (farming, industries, infrastructure), by now down to more or less one third of the working population;
- a parallel shrinkage of manpower supply in the same sector due to the transformation of the population's social structure and to the changed income potentials due to a better-educated work force
- a trend for human and material resources to be concentrated in the production, commercialisation and distribution of knowledge and information-intensive immaterial goods and services (R&D activity, education, assistance, communication and virtual transfer of goods and people, software production);
- a decreased demand for room for production activities, for infrastructure, for buildings, for waste storage, etc., as a result of the availability of

technical alternatives and of consistent program choices;

- a reduced population growth rate, not only in the industrialised countries, but also on a world scale (contrary to all projections). The increasing rates of the eighties (peak rate in 1989: 1.9%) have been replaced by the decreasing rates of the present decade (lowest rate in 1997: 1.4%) that suggest a growth rate equal to zero for the year 2015.

The cultural, economic and social implications of these transformations mostly remain to be explored. Some note the social fragmentation and the consequent decline in social integration characteristic of the post-industrial era, in opposition to the solid class composition in Ford's day; some claim instead a higher degree of relationships and a better quality of social cohesion than the post-industrial era, owing to the quality and quantity of relations due to the spread and use of relational goods (or of proximity).

However the continuous decrease of job utilisation in the production function with materialistic intensity (especially of capital) because of the reduction of job unit numbers necessary to produce a defined volume of production, is compensated by a production function based on technological innovation and information, and also by new types of needs satisfaction, needs that express themselves in a new way.

During the Ford era some kinds of demand were not considered appropriate for the creation of domestic wealth because were satisfied by alternative ways than those of market demand (within the family for example), and so they went out of the production-exchange-consumption circuit[4]. Now in the post-industrial era the same types of demand support a lot of production functions characterised by low capital intensity and a high labour use. Amongst these there are a lot of services for people, and in addition there are cultural services, entertainment and socialisation services, services for enjoying environmental sites, that represent additional sources of well-being and areas for the utilisation of human resources.

These functions are characterised by a low use of real capital and by high utilisation of human capital and natural capital; utility generated not only by the relationship between needs and goods but by that among people or among people and environmental resources; modest capacity for the production of extra-profit a low possibility of standardisation and of efficiency measuring.

We are dealing here with production functions that represent new sources of wealth because they answer in a real way the historical evolution of needs, although they avoid the measuring paradigms that we actually use.

The needs in the Ford era's phase of development were expressed and satisfied at the family level because the family made up the elementary unit of consumption, and the production process of goods for the house and for familiar use (cars, household appliances) was able to meet in a real way the qualitative and quantitative variations in demand.

Now in the post-industrial era these latter needs are largely satisfied, and the family has lost the role as the place and form for the demand for well-being and the organisation of consumption, while at the same time needs arise on a different scale of aggregation that require complex utilities, that is demand on an extended social scale that can be satisfied by the supply not of single goods related to defined types of activity, but through production functions that require system-level policies that involve social organisation, commit the institutional structure, and imply forms of community learning[5]. The result of this latter transformation is that relationships among needs and goods capable of satisfying these are less exhaustive and exclusive, so that the full satisfaction, at least of some needs, requires the involvement of other physical, personal and social relationships. For this reason we generally tend to refer to the post-industrial economy as the economy of relationshps[6].

This means that the dynamics presently underway strongly favour a strategy of sustainable development, especially of tourist activity. In fact, policies and development projects based on labour-intensive investments are no longer advisable; non-environment-friendly productions entailing a large use of materials, energy, water and space are no longer competitive. Therefore, as these activities are not compatible with the environment due to their obsolescence and non-sustainability, they must be abandoned and replaced by strategies of development oriented to devising processes for the production of highly non-material-intensive goods and services which, by their own nature, boast an elevated productivity even in the case of directly productive sectors and, calling for less traditional forms of labour, allow more time to be devoted to recreational and tourist activities.

From these considerations a strong option emerges for the development of tourism in the post-industrial era, an option that leads toward a perspective of sustainable development. By their own structure tourist activities are compatible with the objectives of sustainable development and seem to respond with a certain efficiency to the search for relationships and complex utilities pursued by both consumers and producers (these complex utilities are also prompted by the values attributed to the use of the environment), as they respond efficiently to new needs that cannot be met any longer by a *linear* increase in production and that cannot be well expressed by the traditional production-market-consumption circuit and do not fall within an individual type of activity.

This does not mean that the issue of environmental sustainability of tourist activities is less urgent because it appears encased in the spontaneous dynamics of the post-industrial economy. Quite the contrary: it calls for a greater attention to the elaboration and experimentation of models of development of this kind, as tourism is a complex and delicate sector that activates a multiple sectorial interdependence both upstream and downstream. Therefore the achievement of the sectorial objectives of sustainable development requires the entire economic system to opt comprehensively for a model based on the environmental sustainability of growth.

As a result the adoption of a strategy of sustainable development in the tourist sector calls for the convergence of choices and behaviours not only of those involved in tourist activities, but also of all the other actors in the economic system. This requires the settlement of often conflicting interests and diverging expectations especially at a time of great uncertainty in terms of the choice of the policies likely to promote new forms of *development without* limits, the choice of the models on the basis of which to implement these policies, the choice of the actions to be taken and the structures necessary to pursue objectives which are compatible with both a post-industrial scenario and a perspective of sustainable development which expresses the content and concept of a growth without limits.

3. SOCIAL EXPECTATIONS AND SUSTAINABILITY. A TOURISM DEVELOPMENT PERSPECTIVE

But what could be the effects of a change in the forms and contents of wealth, of a change in the structure of demand and consumption from a sustainable development perspective? And how are all these factors linked to the sustainable development of tourist activities?

First of all, all these kinds of changes are matched by variations in the structure of social wants in terms of their demand for well-being that, albeit apparently widespread in the community, no longer overlap with the theoretical concept of a well-being which is essentially exemplified by the satisfaction of one's needs, i.e. with that indistinct set of flows of utility produced by man in a given period of time by means of processes of production for goods and services. This definition is missing two components. The first is constituted by the set of utility flows generated by the relationships amongst people, the second by the whole set of utilities occurring in nature. Therefore well-being should result from:

a) *produced* flows, in which the role of the real, material capital produced by man prevails, human labour remaining instrumental to this process;
b) *relational* flows, generated by organised relations between and amongst people, in which organisation and labour factors (intangibles) largely prevail over real capital;
c) *natural* flows, where natural resources represent the prevailing factor, labour and relations being instrumental to getting access to these.

For the sake of simplicity, we assume that relational and natural flows represent a subset of environmental well-being, i.e. the aggregate of utilities offered (as opposed to the ones produced) by the different kinds of capital available in nature or in the relational system. Post-industrial development is characterised by the growing role of the latter and by the transformation of the structure of produced well-being, which is in fact produced by assigning a major role to *intangible* capital. This intangible capital is accounted for by man's knowledge and skills and tends to replace material capital.

Sustainable development may be considered as the target of this overall dynamic process, the result of the processes by which material capital is replaced by intangible capital and produced well-being tends to be replaced by environmental well-being to create a comprehensive social behaviour. It may be observed that post-industrial development and sustainable development partially coincide, but the latter cannot do without the former, as the need for sustainability remains linked to a continuous substitution of natural resources with man-made resources; therefore the more non-material the content of the latter is, the higher the likelihood of sustainability.

The ideal type of sustainable development should occur when the growth of the second aggregate is such as to compensate for the decrease in the first aggregate with a residual, or at least null, net margin. Nevertheless this condition of equilibrium amongst produced, and nature and relation-derived well-being represents a tendency for a number of reasons.

First of all because this evolution is hampered by the existence of informational asymmetries due to the fact that the market is not prepared to give information on these types of demand – relational and environmental goods/services – or does not take into account these areas of interests. Therefore the decisions that have an impact on the creation of well-being do not find any applicability on the market that, on the contrary, provides those kinds of information that favours the emergence of individual expectations diverging from community ones and contrasting with the above equilibrium tendency. Given the above asymmetries, the operators tend to project their expectations for well-being in the present time, thus making choices favourable to a rapid evolution leading to a general social well-being, an objective which may be pursued only through a rapid dynamics that is incompatible with a state of equilibrium that relies on an overlapping economic and environmental dynamics. In fact, as is known, the produced flows, which represent the economic system, and the natural flows, which exemplify the environmental one, rely on completely different dynamics, as the former are rapid and the latter are slow. Conversely, long-standing relationships exist between produced resources, whereas amongst natural resources and man and among the factors that produce relational well-being short, contingent relations are established based on personal relations. In the former case there are, in fact, hierarchical, dominance/dependence relations, whereas in the second case mutual relations are at play.

If it is true that the desired equilibrium still remains in the form of a trend, a sustainable development strategy could be feasible if, on the one hand, a threshold of total well-being is guaranteed establishing a cut-off, i.e. a point below which development would not be possible (a possibility determined by a positive variation in the natural and relational flows of utility not sufficient to compensate for the negative variation of the produced flows), and a threshold above which sustainability would not be possible (a feasible hypothesis, should the increase in total well-being be determined by a positive variation in the produced flows due to the use of partially compensated environmental flows).

As we cannot envisage an ideal state of equilibrium, this hypothesis could lead to various suboptimal situations all circumscribed by the development threshold on the one hand, and on the sustainability threshold on the other.

It is in this area that we find the objectives and the real potentials in terms of development without limits (in the above sense) and it is towards these objectives and potentials that the economic policies for a paradigm of sustainable development should be oriented. Nevertheless, as we lack a theory of natural capital the problem remains to detect those thresholds of development and sustainability that define the evolution of general social well-being, i.e. the marginal rates at which real capital, intangible and natural capital can substitute each other. These transformations would modify the structure of social expectations and interests represented on the market as they would give visibility to the new demands for well-being felt by the whole community, which would be in contrast with the individual positions regarding profits and group interests clearly represented by the market.

Once again the issue of sustainable development emerges as a cause for social disaggregation and the underlying conflict between supporters of individual interests and supporters of community interests. Therefore the operational models implying the adoption of marginal rates of substitution of material capital for environmental capital, with all the relevant consequences, need to be planned and turned into programs and instruments that meet operational choices.

Tourism is a complex sector also because, in order to fulfil its production function, it needs all the different kinds of capital: real, intangible and natural. In fact, it utilises both product and environment utility flows thus contributing to the creation of each one of the aggregates of social well-being. This means that tourist activities utilise and produce market goods/services for which one has to pay and, at the same time, confine themselves to also utilise (but not produce) free (environmental) goods/services for which one does not have to pay.

Therefore, in the process involving the transformation of natural flows into utility product flows, tourist activities enjoy positive externalities. But the surplus produced by the transformation processes in the tourist sector is only partially due to the use of manufactured goods and manpower, deriving mostly from the natural utility flows. Nevertheless, as the latter do not have a well defined marginal productivity it is the marginal productivity of the manufactured goods that prevail, thus justifying the continuous passage from free goods to market or manufactured goods. This logic applies the rules regarding the function of the manufactured goods to the function of tourist activities, and leads to the over-exploitation of the processes of transformation of natural goods into market goods/services, since the former allow for the maximization of the weighted marginal utility/productivity and to obtain extra-profits.

In fact, a rational behaviour based on the rules of economic theory would call for the maximisation of the weighted marginal utility/productivity of the goods and services paid so as to obtain an equal weighted marginal utility/productivity of the factors employed. But as natural goods have no price their weighted marginal utility/productivity remains undetermined and therefore they cannot be included in the calculation. It is for this reason that the production process for tourist goods/services should annul the weighted marginal utility/productivity of the factors employed in terms of their market price. This latter, however, does not comprise the cost of natural resources; therefore a marginal utility/productivity which is null from the individual point of view should correspond to a negative marginal utility/productivity for the community.

It is at this point that a clear conflict emerges between social and either individual or group expectations, since the former claim to avail themselves of free goods not only in the present but also in the future and, for this reason, to regulate the transformation processes of these goods into market goods/services, to get access to the enjoyment of the same natural goods according to well-defined rules, and to participate in any extra-profits produced so as to implement one of the sub-optimal well-being situations. On the contrary, individual or group expectations are focused on short-term well-being objectives, on erecting barriers to access, and on getting the entire extra-profits. These two kinds of expectations are so clashing that conflicts are inevitable. It is for this reason that it is so important to agree on the terms of utilisation, i.e. on the regulating mechanisms that have to supplement the market ones by integrating functional information with the production of regulatory information. Sustainable development policies are given the task of pursuing this result by *concerting* apparently incompatible expectations.

4. TOURISM GLOBALISATION AND LOCAL SUSTAINABILITY

So far we have not tackled the problem of the implementation scale of sustainable development. Nevertheless, once general rules and macro objectives have been globally established, whether or not this scale should be local does not seem to be in question. The local scale seems in fact the most adequate to elaborate, propose and apply the instructions for the use of material, immaterial and natural capital that, due to the absence of well-defined substitution parameters, is determined by information symmetries and by the creation of a structure of social expectations by means of non-market mechanisms. In order to be efficient, these mechanisms must be able to assess correctly both the quantity and the quality of the available information and the well-being expectations present in the community. This cognitive process can only occur on a local scale, since both information and expectations are expressed by the specific characteristics of the local economy, territory and environment. One could therefore say that there exists more than one model of sustainability, i.e. there are different avenues leading to sustainable development depending on the features of the local social, economic and environmental context.

In the light of this evidence the global approach to sustainable development has clearly shown its limits. This is due to a series of reasons. First of all the theorem of the globalisation of development policies and of the regulations inspired by the principles of environmental sustainability have long been exploited as an alibi to block the enforcement of prescriptions in this field, along with the launching and development of global programs of post-industrial development meant to favour sustainable development, with the obvious objection that the exercise of control should not be possible without the consensus of all the components of the anthropic system, but in reality because the globalisation of strategies of this kind calls for the transfer to developing countries of the technologies necessary to transform forms of development of the industrial type into models of the post-industrial economy. In other words, the globalisation of development sustainability has been long adopted by the anthropic system to keep the power of action in a climate of complete anarchy in the field of the environment, following the "policy *of the nature of things"* on the road to globalization.[8]

In addition, on the global scale it is comparatively easier to make concerted decisions in the field of the environment as there are few, well-known decision-makers, which generally coincide with sovereign states, and also because the structure of social expectations and the interests at stake are always considered at the level of large clusters. It is certainly more difficult to implement efficient operational policies and programs, as the global view is not objectively able to assess the extent of resources, the modalities of managing them, the structural and technological characteristics of the resulting production relationships, the needs for adjusting the development strategies to the local scale in terms of the objectives of sustainable development, i.e. transferring general principles from the political regulatory level to the operational one, clashes with the various general situations with the resulting creation of conflicts in terms of needs, values and interests due to the presence of different levels of development, societal structure, expectations, human skills, know-how, technology and peculiarities of the resources existing in the different territorial contexts.

Moreover the environmental system is as different in its local components as the anthropic system. Therefore the same relationships between economy and environment greatly differ in the local context, thus contributing to create the local dimension for development and for the anthropic and environmental systems without being for that a portion of one or the other system, since the local contexts are characterised by a remarkable variety of endogenous resources.[9]

This means that a credible operational approach must start from the origin of the production processes in order to determine a policy which is shared and which orients the actions meant to achieve production and employment objectives towards a context of sustainability meant as a value added to the production processes and to the products (as an expression of local variability) and not only as a prescription to pursue the above objectives.

In addition the territorial variability of productive and social relations and of the expectations that determine them can only be verified on the local scale, where it also possible to verify the links amongst production processes and local resources as well as amongst the same resources and the different groups of interests that characterise the identity of the system. As these factors vary from one local system to the other, so the identity of each and every local system changes and with it also the operational interpretation of the sustainability of development. This widespread atmosphere of transformation seems to invest also tourism which is more and more open to the global scale, and maybe just for this reason induces the local communities to compare its fallout to their expectations. In fact, the globalisation process regarding the services instrumental to tourist activities (marketing, commercialisation, transportation) has weakened both the control and social regulation of the latter, since the production activities of the tourist services still remain located inside the incoming areas of the tourist flows, whereas the instrumental activities are more and more based in those countries these flows depart from.

Apart from the undeniable effect of such labour sharing, this polarisation results in an unequal distribution of costs and profits, as it tends to attribute the former to the recipient areas and to share the latter between the recipient countries and those that emit tourist flows, thus favouring the latter because it is in these countries that the immaterial processes with a high welfare producing capacity are located, i.e. the added-value processes for each product unit.[10]

In fact, in addition to direct production costs, recipient areas are subjected to the externalities of the production and consumption activities. These externalities are not perceived by the market, but they are "felt" by the community. It is true that these may be either negative or positive, but in view of the heterogeneity of the tourist products, it is necessary to define their source and their production process in order to characterise them. If positive externalities were induced by processes involving the transformation of goods into material products, their cost-benefit should be evaluated as well as their effect on the structure of the social well-being, so as to establish the correct balance between the externalities. As it is the local system that bears the costs, negative externalities, profits and positive externalities, the expectations for the regulation of well-being on the basis of social expectations in terms of sources and alternative means of creation and in terms of a greater present and future well-being are fully justified. Therefore the globalisation of tourist services has reinforced the local scale of the policies meant to regulate them.

In addition, tourism is a consumer of public goods and, therefore, by subtracting resources from alternative uses, it influences the composition of well-being and the creation of the structure of the social expectations in terms of resource allocation. This kind of assessment cannot fail to occur on a local scale in view of information symmetries, the creation of expectations and the elaboration of measures. Should the information asymmetries be abolished in terms of social expectations in tourism, tourist operators would be more willing to pay to safeguard the non-reproducible factors of their own production function, including some natural assets, because they expect the reinforcement of their competitive position to produce an added value likely to recover the cost of negative externalities.

Yet tourist attractiveness often turns into a pre-competitive factor also for the other sectors of the economy that benefit from it without having to bear the relative costs (i.e. they enjoy positive externalities). It is in this way that the value chain of an environmentally sustainable tourist and development process is distributed to all the components of the local system.

As a result there emerges the problem of a fair re-distribution of the costs and benefits of tourism's sustainable development, a possible trigger for an additional conflict amongst the members of society. Therefore the issue of sustainable development confirms that − if ill-coordinated − it is a possible cause of clashes not only between generations, as is generally thought, but also between productive sectors, as is less frequently thought. For this reason, in order to be sufficiently integrated in terms of productive classes and sectors, an operational model of sustainable development should be able to combine the relations and the interests meant to obtain positive externalities with a fair distribution of costs and negative externalities.

In this way the risks of fragmentation and social conflict, specific to the policies of sustainable development, would be minimised.

5. DEVELOPMENT POLICY: ENVIRONMENTAL LINKS AND THE TERRITORIAL PACT

The areas of conflict of a society projected towards sustainable development may be ascribed to two kinds of issues. The first concerns the territorial scale and the level of the authority that is supposed to define the development model, to work out and implement the conditions for capital accumulation and for the use of the environmental resources. The second concerns the intersectorial and the intertemporal allocation of both material and immaterial resources as well as the distribution of costs and negative externalities, and of profits and positive externalities determining the sectorial composition of well-being, and the creation of intertemporal social expectations.

Leaving the settlement of these conflicts to market mechanisms would probably determine a reinforcement of those interests targeted at creating short-term well-being flows. This would also strengthen the trend to transform natural assets into market goods. In addition, this would also hamper the development of entrepreneurial self-employment sectors that, capitalising on interpersonal relations, meet the demands for a modification of the structure of social expectations as well as the demands for new relational goods and services. These demands are not met by the market, as they do not produce extra-profits and do not attract the attention of entrepreneurs.

All this would hinder the evolution of a given system towards the tendential equilibrium between produced well-being and relation - and nature-derived well-being previously referred to as the ideal type of sustainable development in a post-industrial society.

In addition, consider that these conflicts concern the social demands for the definition of the modalities of capital accumulation (in its broad sense that includes material, intangible and natural capital), as well as the demands for establishing the structure of well-being and the options for its intertemporal distribution. What is more, if we consider that conflicts derive from the opposition between individual or group interests and social expectations, it is clear that the settlement of these conflicts must be based on social policies that envisage systemic actions in terms of the restructuring of the whole society, the institutions, and the tools for community learning.

The reasons underlying the above conflicts are particularly strong when we have to face a strategy of sustainable tourist development, as this sector is characterised by:

- a production function based on the use of produced or material capital, relational capital or intangible capital and natural capital that must be regulated by means of extra-market mechanisms, their substitution rates remaining undefined;
- the presence of multiple externalities and sectorial interdependence that encourage the structure of community expectations and stimulate the emergence of new demands for well-being;
- the gradual globalisation of service production processes that determines an international sharing of work and an unequal distribution of well-being between tourist inflow and outflow areas;
- the essential role of public assets, which is linked to the all-important function of the public decision-makers in regulating the modalities of use and the resulting conflicts of interest.

These peculiarities, which are specific to tourism, highlight the need for elaborating an operational planning model that enables us to create a consensus around the objectives of sustainable development by adopting extra-market regulating tools. This planning model should be made up of at least three fundamental components:

- a territory where the environmental system and the community that intends to pursue a sustainable development are both fully embedded;
- a decision-making and planning method that represents and takes into account the interests of the different groups in society according to a principle of mutuality based on both community interests and expectations but also on the technological trajectories of the territorial system in which the environmental system and the community that intends to pursue a sustainable development are both fully rooted;
- a set of indicators of efficiency and efficacy derived, if need be, also from shadow prices or from other parameters of assessment of both positive and negative externalities integrated, on an equal basis, with market prices according to a scheme of financial assessment on the one hand, and economic assessment, on the other.

The necessity of giving an institutional framework to these elements has resulted in the adoption of negotiated planning as a process of decision-making, of concerted action as a methodology to create a consensus and information symmetries, as well as of the territorial pact as a tool for the definition of the conditions of use of the resources and their intersectorial distribution, and hence as a guide to assess community expectations and distribute well-being over time.

In order to build a consensus around this important institutional economic and social innovation, the negotiated planning process adopts measures to regulate transactions and relations which are completely non-aligned to market parameters.

First of all, negotiated planning tries to give visibility to the relations and to the conditions that are at the basis of local development, building schemes agreed upon by both private and public subjects who have never before engaged in dialogue except on a contractual basis, and thus in a climate of confrontation regarding their reciprocal interests.

It is for this reason that negotiated planning is a sort of experimentation in the field of a new model of *governance* with a systems framework. In this bargaining phase, the enterprises interested in increasing their assets will search for short-term profitability, whereas the community and institutional systems will tend to impose a post-industrial, sustainable development perspective, i.e. they will tend to develop the system's "slow resources of competitiveness"[11]. The preference granted to the one or to the other aspiration will constitute the indicator of the social expectations for the intersectorial and intertemporal distribution and composition of well-being.

Negotiated planning makes available a full range of operational tools (program agreements, program contracts, territorial pacts, area contracts). Amongst these tools, those that seem to be more adapted to system actions are the territorial pact and the area contract, which differ in terms of their impact on the creation of community expectations, on social relations and resources. This impact seems, however, to be more pronounced in the territorial pact. This is a typical extramarket regulatory instrument, not only because it partially relies on non-market mechanisms to create information symmetries and to determine community expectations, but also because it tries to develop virtuous circles of local development and efficiency by forcing the conditions of use of the factors as they are dictated by the market.

So far these latter advantages (facilitated credit lines, tax holidays, technical assistance services, flexible jobs, and administrative and bureaucratic simplification) of the territorial pact have been attracting both the attention and the consensus of the operators, but at the same time they have also called for the elaboration of programs for the enlargement of the productive basis and have oriented the local expectations towards objectives of increased material production in exchange for employment, without pre-defining a development option, but rather, encouraging its implicit emergence from the choices made. These latter have been oriented towards a traditional model of development, scarcely open to the dematerialization of the production processes and to the perspectives of sustainable development.

In the light of these first results, can the territorial pact be considered as a tool to implement a sustainable tourist development? And, if yes, with what conditions?

The options of development that emerged from the first implementations of the territorial pact seem not to encourage positive answers. Nevertheless there is no physical, cultural and legal constraint that impedes adapting the tools for negotiated planning to a design for sustainable development, provided that some pre-defined conditions are met.

First of all, we have to recover the role of negotiated planning as a unitary process for the creation of community expectations and decision-making, and as the only site for the creation of consensus and information symmetries, as its development as a system tool (territorial pact and area contract), and a tool supporting the individual actions (program agreement and contract applied also to tourist initiatives) has weakened the original objective of concerted policies that consisted in making the choice of a comprehensive model of local development with the participation of all the components of civil society. As sustainable development cannot do without a systems option, negotiated planning must be considered the most adequate process to make this kind of development feasible. The territorial pact may represent the tool for implementing community expectations in terms of use of resources and in terms of the intertemporal distribution of well-being, provided that this implementation tool does not surreptitiously modify the unitary planning design. This calls for additional conditions dependent on non-local economic policy options and on the choices of the operational criteria specific to negotiated planning.

Among the former it is essential that the territorial pact also be open to tourist activities. Presently all the sectors of the economy can get access to the territorial pact with the exception of tourism, even if it is by now recognised that tourist activities greatly contribute to the accumulation of wealth, job creation, and the implementation of strategies for post-industrial sustainable development. In addition, to get access to the privileged conditions of negotiated planning, the local communities are asked to define and make clear the choice of their development model and the reasons for this choice in order to be assigned extra-market conditions for the use of resources. These resources are selected and granted inasmuch as they comply with community expectations and with the model of a post-industrial sustainable development.[12]

This means that the territorial pact cannot be generalised with regard to local contexts that make of its variety their strength. In addition, the territorial pact cannot be selected only according to the kind of action that has to be carried out but also in terms of its actual positive contribution to sustainable post-industrial development. From this point of view the present evaluation is made in negative terms, i.e. sound territorial pacts cannot have a negative impact on the environment, but any added contribution on their part to the system's options is completely disregarded. This reiterates, without eliminating them, the causes of conflict and, at the same time, prevents the achievement of quality employment objectives, given that quantity is a highly changing concept in modern economic dynamics, especially when it is associated with the materialisation of production and the levels of development.

The lack of a real selection is producing paradoxical situations, including the overlapping of exceptional instruments of investment financing, the claim to correct local situations of environmental crises by encouraging the localization of real fixed capital with poor, if any, effects on the integration of the processes involving the production of material goods with those involving the production of relational and immaterial goods and services. Finally, the absence of a delimitation of the territorial pact as a tool of extraordinary intervention is producing expectations of facilitated credit lines and simplified administrative procedures, i.e. of *deregulation,* which are incompatible with the social regulation of local development and with the objectives of sustainable development.

This deficit in terms of a planning approach leads to a strengthening of community expectations for a traditional development model. Experience is teaching us that, in the absence of pre-established conditions for the use of resources and of a conclusive option of sustainable development, the actual choices of those regions that sometimes prioritise satisfaction of their basic *needs,* including employment in economically depressed areas of developed countries, do not follow a path of sustainable development but, quite the contrary, show their growing availability to exchange free goods for market goods. This means that they prefer to postpone dealing with the issue of the cost-opportunities of their options and to renounce future well-being as well as the system's competitive resources.

6. CONCLUSIONS

The local-scale organisation of community expectations in an institutionalised form in the economically depressed regions of the industrialised countries tends to encourage the enlargement of the productive basis of material goods. Therefore, proposing options involving the generalised dematerialization of the productive system according to strategies of industrial development and towards objectives of sustainable development turns out to be utopian and likely to bring about social conflicts.

Nevertheless it is also true that, together with the demands for meeting basic needs such as employment, other social demands are emerging in these regions which are directed to defining the ways to accumulate and use capital - be it material, intangible or natural - aligned with the forms of post-development adopted in developed regions, and also directed to determining the ensuing composition of well-being as well as its distribution over time.

From the operational point of view a strategy of post-industrial development is preparatory to the launching of policies of intermediate sustainability, since it modifies the structure of community expectations and encourages the emergence of new demands for well-being.

The co-existence of apparently incompatible community expectations is suggestive of underlying conflicts and the confrontation of interests whose settlement and prevention should take place through social policies that envisage *systemic actions* for restructuring society, institutions, and tools for community learning. These processes must adjust themselves to the different kinds of societies, economies and local environments. Therefore the territorialisation of a model of sustainable development is a pre-requisite called for by the necessity of defining in concrete terms the relationships existing between anthropic and environmental systems and of defining the quantity, focus, and quality of these relationships. This means physically and culturally defining the operators of production, the consumers, the institutions and the other members of society, as well as their needs and expectations, their strength, and finally, their availability to accept or not to accept the prospects of sustainable development.

Tourism seems to be the most suitable sector for launching a model of post-industrial sustainable development as, even if opposing production functions and interests are at work in it, it lends itself to organising local interests around the objectives of the dematerialization of the processes of service production and of sustainable development, also in view of the role played by relationships and environmental resources in the composition of the tourist supply.

In spite of these potentials, the serious problems linked to its structure, and its adequacy to promote systems policies, tourism has still not taken on a role appropriate to the wealth it produces in local development policies as well as in the tools for consensus formation for these same policies and actions. It seems legitimate to suspect that, since sustainable tourist development imposes some constraints on the entire local economic system, thus conditioning the development of other productive sectors, the access to other regulatory procedures such as negotiated planning, concerted action, and to extra-market mechanisms for the use of resources, like those envisaged by the territorial pact, is deliberately hindered, leaving open access to non- system measures directed at supporting timely initiatives, which are distant from the objectives of dematerialization and of sustainable development, such as the various laws granting financial incentives to structural investments and some instruments of negotiated planning, including program agreements or contracts.

Precisely this resistance to extend the territorial pact to tourism shows, on the contrary, that the tourist sector is the most appropriate for representing community expectations in terms of the objectives of post-industrial development, as well as in terms of its potential impact on material forms of development and on those *equilibria* based on the convergence of local profits and extra-local interests derived from the transformation of natural goods into market goods.

NOTES

1.   G. Ruffolo (1995), pp. 9-11
2.   M. Bresso (1993), pp. 78-81
3.   For a critical review of the first investigations on the post-industrial economy, see B. Frankel (1987), pp. 24-25
4.   G. Fuà (1993), pp. 47-48

5. A. Montebugnoli (1997), pp. 16-20
6. De Vincenti and A. Montebugnoli (1997)
7. M. Bresso (1993), p. 71
8. U. Beck (1999), p.41
9. G. Becattini and E. Rullani (1993), pp. 29-30
10. F. Delbono and G. Fiorentini (1996), pp. 128-137
11. A. Bagnasco (1996), pp. 4-9
12. E. Gerelli (1995), pp. 209-223.

REFERENCES

Bagnasco A. (1996), Governare i tempi dell'economia, *Impresa e Stato*, 35: 4-9.

Becattini G. and E. Rullani (1993), "Sistema locale e mercato globale", *Economia e Politica Industriale*, 80: 29-30.

Beck U. (1999), *Che cos'è la globalizzazione*, Carocci, Roma.

Bresso M. (1993), *Per un'economia ecologica*, La Nuova Italia Scientifica, Roma.

De Vincenti C. and Montebugnoli A. (1997), *Introduzione*, in C. De Vincenti and A. Montebugnoli (eds.), *L'economia delle relazioni*, Laterza, Milano.

Delbono F. and Fiorentini G. (1987), *Economia del turismo*, La Nuova Italia Scientifica, Roma.

Frankel B. (1987), *The Post-Industrial Utopians*, Polity Press & Basil Blackwell.

Fuà G. (1993), *Crescita economica*, Il Mulino, Bologna, pp. 47-48.

Gerelli E. (1995), *Le società post-industriali e l'ambiente*, in A. Quadrio Curzio and R. Zoboli (eds.), *Ambiente e dinamica globale*, Il Mulino, Bologna, pp. 209-223.

Montebugnoli A. (1997), "Il paradigma post industriale", in C. De Vincenti and A. Montebugnoli (eds.), *L'economia delle relazioni*, Laterza, Milano.

Ruffolo G. (1995), "Dallo sviluppo della potenza allo sviluppo della coscienza", *Geotema*, 3, pp. 9-11.

11. THE IMPACT OF CLIMATE CHANGE ON FLOWS OF BRITISH TOURISTS

DAVID MADDISON[*]

1. INTRODUCTION

Insofar as British people are now, largely for the purposes of recreation, spending more time away from home than ever before the climate of other countries may be important to the welfare of the British - if only for purely selfish reasons[1]. Organised trips are now available taking people from Britain to destinations in North America, Asia, Africa and Oceania whilst package tours to the Mediterranean are now almost a quintessential part of British life. The rapid growth in international tourism whether as a part of an organised tour or independently is a reflection of greater leisure time (due in part to an ageing population) and a growth in real incomes. There is also evidence linking the growth of package holidays in the Mediterranean with reductions in cost caused by more fuel efficient planes and with a decrease in the cost of accommodation (Perry and Ashton, 1994). In addition to the relative price of different locations choice of destination is presumably influenced by a desire to visit particular landscapes or sandy beaches for recreational purposes, motivated by a desire to explore or renew cultural ties between countries or to partake of the alleged health benefits of particular locations. Poor health has often been cited as a reason for making a journey (e.g. the remedial properties of hot spas, mountain air and coastal climates). Even until quite recently a tan was considered rather a "healthy" thing to possess. Choice of destination is also heavily influenced by the image that a country has with regards to its political stability and crime rate (e.g. the recent poor publicity surrounding Florida following the murder of several British tourists).

But a major factor in choice of both destination and time of departure is climate. Indeed, when British tourists go abroad they are often described as being "in search of warmer climates". Tourists might be construed as making a decision to go abroad in order to gain some short-term climatic advantage. Certainly, with regard to domestic tourism retired people in particular can be observed migrating south for the winter in America to Mexico whilst in Australia they head north to the "Gold Coast" resorts of Queensland. Both "push" and "pull" factors are clearly at work. Whilst the importance of climate to both domestic and international tourism is hardly disputed there are not very many empirical studies which have explored the implications of various climate change scenarios for international tourism and welfare (for an exception see Wall, 1992).

The impact of climate change on tourism in Britain is dealt with by Smith (1991) and published in the Government's own review of the potential impacts of climate change on the United Kingdom. According to Smith, the tourist season may lengthen and tourist satisfaction may increase. But here too there is no attempt to determine the changes in the overseas destinations of UK tourists, changes in the number of holidays taken (although these change are argued to be small when compared to likely changes arising from greater leisure time and higher incomes) nor to value the costs or benefits from changes in the climate insofar as international tourism is concerned. The purpose of this paper then is to assess in a quantitative fashion, probably for the first time, the importance of climate as a determinant of choice of travel destination for British residents among a number of other possible factors including travel cost.

From a purely strategic perspective it is of obvious interest to examine how the numbers visiting different sites change as the climate changes. Many island economies in particular are heavily dependent on tourism and if climate is what tourists are seeking then climate change may have significant consequences for these island economies. Furthermore, following the methodology outlined in this paper it is possible to compute a money-metric measure of how welfare changes as the attributes of a set of sites change (the welfare of tourists changes in the sense that more desirable climates may be brought closer to home). There is also a possibility that several low-lying island states may well become "unavailable" in the sense that they risk being inundated by rising sea levels. The methodology employed enables monetary values to be placed on this eventuality at least in so far as "use" values are concerned. The methodology rests on the fact that different sites are characterised by different travel costs and accommodation costs. By observing differences in visitation rates it is possible to examine the rate at which individuals are willing to trade off higher money costs against desirable "site attributes" such as climate. The economic values of changes in both site quality and availability may be of interest to those seeking to compile an overall damage cost assessment to the effects of climate change.

The remainder of this paper is organised as follows. The next section examines the plausibility of the one of the key assumptions underlying revealed preference techniques: the existence of perfect information. The third section describes the model used to explain the pattern of observed visitation rates, the extraction of money metric measures regarding changes in site quality and availability and discusses some alternative specifications. The fourth section looks at the data sources available for the purposes of estimating the proposed model. The fifth section considers the results of the statistical exercise. The penultimate section uses the results of the exercise to illustrate the impact of climate change on particular tourist destinations and the final section concludes.

2. CLIMATE AND TOURISM: THE ASSUMPTION OF PERFECT INFORMATION

The first information that potential holidaymakers encounter regarding the climate of a particular destination is through the travel company's brochure. The image of an attractive climate is cultivated in the mind of the consumer (usually) bounded by the requirement to remain within certain standards. There is obvious scope to be selective in the presentation of particular climate variables. The existence of warm temperatures in the Mediterranean during wintertime is pushed but the relatively higher rainfall (compared to London) occurring at that time is downplayed. Holiday tour operators frequently use the same "blue sky" photographs in their summer and winter brochures. Even when climatic tables are given there is a concern that these raw data are not readily understood by the potential tourist and that some "expert interpretation" is required.

The view that tourists were largely ignorant of or were purposefully deceived about the climate of their intended destinations (and therefore suffered disappointment as a result) and were moreover incapable of assessing for themselves raw climate data led to the construction of indices which purported to objectively evaluate tourist potential of the climates of different countries. In the view of the contributors to this literature the climatic resources of the world were not being fully utilised or, to put it another way, the inadequate provision of information resulted in market failure. In these indices different components of climate were subjectively weighted and placed on a labelled scale.

Perhaps as a result of the view that tourists do not have access to perfect information there are no examples of taking a revealed preference approach to tourist flows in the literature. But are tourists really so lacking in information? Do they allow themselves to be persuaded by blue-sky photographs? Tourists presumably have come to expect blue-sky photographs and discount such tricks. Furthermore the traveller has ready access to sources of high quality low cost information which is independent of the travel company he goes with in the form of countless travel guides, television programmes such as The Travel Show, daily weather reports for world capitals published in the newspaper as well as television and radio weather forecasts which now cater for the overseas traveller and even quite specialised weather guides such as Pearce and Smith (1993). The main source of information however is surely from people who have already visited a particular destination. They know how the climate actually felt and can describe it in terms familiar to them. Tourists are also able to take legal redress against tour operators whose holidays were for one reason or another sub-standard. Since the cost of obtaining this information is low relative to the cost of a typical package holiday the tourist has every incentive to take advantage of it (Perry, 1993). It is frankly hard then to conceive of how travel companies could consistently mislead the majority tourists with respect to what type of climate to expect and the concerns of earlier researchers appear misplaced. As such the revealed preference approach, which rests on the assumption of perfect knowledge concerning the attributes of different destinations, can be expected to work well whereas much of the earlier literature looks like an anachronism. Furthermore, the revealed preference approach is capable of expressing in monetary terms the extent to which changes in the climate of different holiday destinations changes welfare which is a fundamental objective of this chapter.

3. THE POOLED TRAVEL COST MODEL

The paper now turns to a theoretical model of the allocation of time and money spent visiting different destinations (or "sites" as they are called in the travel cost literature) and the consumption of other goods. The model employed here is based largely on McConnell (1983) and Johansson (1987). Note that unlike in the majority of the travel cost literature the time spent at particular sites is modelled as being a variable of choice.

Assume that the individual derives utility from the number of visits to different destinations, the time spent on each visit, as well as from the consumption of a vector of other goods. The individual's utility function can be written as:

$$u = u(q, x, t, z)$$

where u is utility, q is a vector of consumption goods, x is a vector containing the number of visits made to each site, t is the time spent on each visit to site j and z is a vector of site quality. The constraint attached to the choices made is as follows:

$$q\,p_q + xp = m + l\,p_w - t\,p_w - ax\,p_w$$

where p_q is a vector of prices, p is the (ticket) price of travel, m is unearned income, l is the amount of time available for work, p_w is the wage rate and "a" is the time required to visit a site. Thus the final term of the right-hand side of the equation represents the economic cost of time spent travelling. Associated with the solution to this problem of constrained maximisation is the indirect utility function V (in which some constants are suppressed in order to simplify the notation): $V = V(p, z)$

Employing Roy's theorem yields a set of demand equations for among other things the number of visits to each site:

$$\frac{V_p}{\lambda} = -x(p, z)$$

where λ is the marginal utility of money (which is treated as a constant). Let p^0 represent the current price of travel and p^c a price so high that no trips are taken at all (p^c may be infinity and is often referred to as the "choke" price). Integrating both sides with respect to p between the limits of p^c and p^0 gives the Consumer Surplus (CS) obtained from the site:

$$CS = \frac{V(p^c, z^0)}{\lambda} - \frac{V(p^0, z^0)}{\lambda} = -\int_{p^0}^{p^c} x(p, z^0)dp$$

Next, differentiating both sides with respect to z gives:

$$\frac{V_z(p^c, z^0)}{\lambda} - \frac{V_z(p^0, z^0)}{\lambda} = -\int_{p^0}^{p^c} x_z(p, z^0)dp$$

But given the assumption of weak complementarity between x and z (see Freeman, 1993):

$$\frac{V_z(p^c, z^0)}{\lambda} = 0$$

In other words, if it can be assumed that there exists a price so high at which the number of trips taken to the site falls to zero, then changes in the level of the site attribute z do not affect utility. Integration of this equation with respect to z gives the change in CS following a change in the level of site attribute z:

$$\Delta CS = -\int_{p_0}^{p_c} [x(p, z^0) - x(p, z^1)]\, dp$$

where z^0 and z^1 are the pre and post change level of site attribute. Note that even though several commodities may exhibit weak complementarity with environmental quality (e.g. flight costs, accommodation costs etc.) all that is required in order to measure the value of changes in environmental quality is the demand curve for one of those commodities.

When many alternative sites are being studied there may be substantial variation in qualities across sites. But whilst travel costs to the same site may or may not differ between individuals (in the empirical application of the model described below they do not) the site quality is the same for everyone. Therefore all empirical models which attempt to incorporate site quality have involved some kind of simplification and as a consequence suffer limitations in their ability to characterise recreation demand accurately (Freeman, 1993). One approach (see for example Smith et al., 1986 or Caulkins et al., 1986) has been to pool all of the observed visitation rates for the different sites and to estimate a single demand function in which the observed visitation rates are solely a function of the own price and quality variables:

$$x_j = x(p_j, z_j) \ \forall j$$

This model is referred to as the Pooled Travel Cost Model (PTCM) and is the model employed in this paper. The fundamental weakness of the PTCM is that it predicts changes in the overall number of visits to a group of sites but does not allow for a reallocation of visits between different sites following a change in the price or quality attributes of alternatives. Moreover it assumes that the coefficients on the own price and quality variables are the same across all sites. By contrast, reallocation effects are dealt with explicitly by the Random Utility Model (RUM) of choice approach to valuing site attributes[2].

4. DATA AND SPECIFICATION

Having outlined its theoretical underpinnings, this section describes the data sources for the variables used to estimate the PTCM. There is little in the way of existing literature to act as a guide to the appropriate specification of the model (at least in terms of what quality attributes are the important ones). Accordingly the analysis should be looked upon as a probationary one.

For the dependent variable quarterly data on international travel by British residents is taken from the International Passenger Survey (IPS) for 1994. Visits abroad for reasons other than holidaymaking (e.g. business trips) are excluded since they are not as responsive to climatic factors. The data set also contains the average return fare paid per person to each destination, average spending on items other than fares and the average duration of the stay. From the latter two variables it is possible to determine daily expenditure. Whilst this is not the same thing as having a sterling price index for the cost of living relevant to tourists it is quite clear that things such as accommodation costs are an important consideration to the potential tourist and also that some countries are considerably more expensive to stay in than others. It was argued in the theoretical model that the amount of time required to travel to the country of interest had an opportunity cost attached to it such that other things being equal nearby resorts are preferred. The average time spent in transit is not available so as a proxy the 'great circles' distance from London is used instead. The great circles distance is the shortest distance which an aircraft could fly to reach a particular destination. Ordinary rectangular (mercator projection) maps found in most atlases obviously cannot be used to measure the great circles distance and instead an azimuthal equidistant projection map is required[3].

GDP per capita in US dollars converted using purchasing power parity exchange rates is taken from the UNDP (1995). Its inclusion in the data set reflects the belief that countries with higher GDP possess better tourist infrastructure (hotels, restaurants, visitor centres etc.). Furthermore some tourists might be upset by visions of poverty and squalor which would greet them in many low income countries. Population and population density are taken from the Times Atlas (1992). Population proxies for the quantity of what might be called the "cultural capital" that a particular country possesses (e.g. notable museums, sites of historical significance, buildings of architectural interest). Population density proxies for what might be referred to as "natural capital" (e.g. unspoilt areas, environmental quality). These are unashamedly broad terms. It is anticipated that whereas the former will be positively related to tourism flows, the latter will negatively affect them.

The attraction of some countries clearly lies in the fact that they possess unspoilt sandy beaches fit for recreation. The total length of beaches found in different countries is available from a report by Delft Hydraulics (1990) and is added to the data set. Climate variables are taken from Pearce and Smith (1994). The climate of the country's capital city is taken since this is arguably the most relevant for tourists (although there are arguments for producing a weighted average of several records to represent the climate of the larger climatically more diverse countries). Two variables are included as a description of the climate: averaged maximum daytime temperature and precipitation on a quarterly basis. The former is included in both a linear and quadratic fashion. Including both linear and quadratic terms allows temperature to exert both a positive and negative influence on visitation rates depending upon the current temperature. Finally, three dummy variables are included to represent the different quarters. The role of these variables is to demonstrate that differences in visitation rates can ascribed to climate rather than any other seasonal factors such as statutory holidays. Among the many other variables which might be expected to have an important influence on holiday destinations but which are not included in the data set foremost among these are variables related to the degree of personal safety. This might in some future study be satisfactorily proxied by the inclusion of countries' respective murder rates. Sunshine too is omitted since it is not collected on a consistent basis for very many capitals. This is unfortunate in that many tourist destinations (such as Cyprus) are renown for their sunny climate.

In total 305 complete observations are available from 88 different countries. The variables contained in the data set are listed in Table 11.1, and the characteristics of the data set are examined in Table 11.2. The different countries represented in the data set are listed in Table 11.3.

Turning now to the functional specification of the model, the demand equation for the PTCM is modelled as:

$$\frac{VISITS_j^\lambda - 1}{\lambda} = \alpha_0 + \beta_1 FARE_j + \beta_2 GDP_j + \beta_3 POP_j + \beta_4 POPDEN_j +$$

$$\beta_5 BEACH_j + \beta_6 PDAY_j + \beta_7 DIST_j + \beta_8 TEMP_j + \beta_9 TEMP_j^2 +$$

$$\beta_{10} PRECIP_j + \beta_{11} Q1_j + \beta_{12} Q2_j + \beta_{13} Q3_j + e_j$$

where the subscript j refers to each different observation in the data set. Two special cases were considered: $\lambda = 1$ and $\lambda \to 0$. These refer to the linear and semi-log models respectively[4].

Table 11.1. Definition of Variables Contained in the Travel Cost Data Set

Variable	Definition
VISITS	Number of visits from the UK
FARE	Average cost of a return fare (£s)
GDP	GDP per capita (1992 USD)
POP	Population
POPDEN	Population density (persons per km^2)
BEACH	Beach length (km)
PDAY	Cost of an extra day's stay (£s)
DIST	Great circles distance from London to the capital (miles)
TEMP	Quarterly averaged maximum daytime temperature of the capital city (°C)
PRECIP	Quarterly precipitation in the capital city (mm)
Q1	Takes the value unity for the first quarter, zero otherwise
Q2	Takes the value unity for the second quarter, zero otherwise
Q3	Takes the value unity for the third quarter, zero otherwise

Source: See text.

In the linear model the impact on visitation rates of a change in the level of any variable is independent of the level of any other variable whereas in the somewhat more plausible semi-log model this is not the case. β_1, being the coefficient on the own-price variable is expected to be negative, as are the coefficients β_4, β_6, β_7, β_9 and β_{10}. In contrast the coefficients β_2, β_3, β_5 and β_8 are expected to be positive. There are no prior expectations regarding the sign of the coefficients β_{11}, β_{12} and β_{13}.

Table 11.2. Characteristics of the Travel Cost Data Set (Number of observations = 305)

Variable	Mean	Std. Dev.	Minimum	Maximum
VISIT	85134.	0.26600E+06	235.0	0.2332E+07
FARE	225.58	148.47	17.00	818.0
GDP	10015.	6950.7	620.0	0.2376E+05
POP	0.4948E+08	0.1570E+09	6700.	0.1100E+10
POPDEN	395.83	1378.5	0.2455E-01	0.135E+05
BEACH	131.03	368.09	0.0000	2970.
PDAY	37.397	15.774	9.287	181.0
DIST	3565.9	2523.3	199.0	0.1168E+05
TEMP	22.496	8.5632	-3.667	38.67
PRECIP	83.404	69.773	0.0000	403.3
Q1	0.23279	0.42330	0.0000	1.000
Q2	0.25902	0.43881	0.0000	1.000
Q3	0.26557	0.44236	0.0000	1.000

Source: See text.

Table 11.3. Countries included in the travel cost data set

Anguilla	Columbia	Grenada	Mexico	South Africa
Antigua	Cuba	Hong Kong	Monaco	Spain
Argentina	Cyprus	Hungary	Marocco	Sri Lanka
Australia	Denmark	Iceland	Nepal	St Lucia
Austria	Djibouti	India	Netherlands	Sweden
Azores/	Dominican	Indonesia	New Caledonia	Switzerland
Madeira	Republic			
Bahamas	Ecuador	Iran	New Zealand	Syria
Barbados	Egypt	Israel	Norway	Tanzania
Belgium	Fiji	Italy	Pakistan	Thailand
Bermuda	Finland	Jamaica	Philippines	Trinidad
Bolivia	France	Japan	Poland	Tunisia
Brazil	French	Jordan	Portugal	Turkey
	Polynesia			
Brunei	Gabon	Kenya	Puerto Rica	UAE
Canada	Gambia	Lebanon	Romania	Uganda
Canaries	Germany	Luxembourg	South Korea	USA
Cayman Islands	Gibraltar	Malaysia	Seychelles	Venezuela
Chile	Greece	Malta	Singapore	
China	Greenland	Mauritius	Slovenia	

5. RESULT

Using the method described by Maddala (1977) it was found that the semi-log model (corresponding to the case $\lambda \to 0$) was indeed most likely to have generated the observed data. The results of the semi-log regression analysis are displayed in Table 11.4.

Overall the regression is highly significant and manages to explain almost 50% of the variation in the log of observed visitation rates. Furthermore the hypotheses put forward in the preceding section are all upheld. Nonetheless not all visitation rates are well predicted which is unsurprising given the importance of country-specific factors in determining choice of destination. As expected, the coefficient on the own-price variable "FARE" is negative and highly significant indicating that, other things being equal, more expensive destinations generate fewer trips. It is also observed that countries with a higher GDP per capita are likely to generate more trips as are more populous countries but that countries with a lower population density are preferred. Countries with greater numbers of beaches are well-liked. The variable indicating the cost per day of visiting the different sites is negative and highly significant indicating that the more expensive a country is to stay in, the more infrequently it is visited. The variable describing the great circles distance from London has the correct sign but has only marginal significance. In part this may be because of the relatively high correlation between distance travelled and fare price[5].

Turning to the climate variables, the variables describing quarterly averaged maximum daytime temperature the coefficients on the linear and quadratic terms are positive and negative respectively pointing to the existence of an "optimal" maximum daytime temperature for tourism of around 29°C. Precipitation on the other hand has a negative coefficient indicating that greater rainfall deters tourists although not significantly so. This suggests that perhaps a different measurement concept other than precipitation such as rain-days might have been more appropriate or else that other omitted climate variables like hours of sunshine have biased the coefficient on rainfall. None of the dummy variables describing the time of departure are significant implying that it is climate rather than other seasonal factors which explain observed visitation rates.

Table 11.4. The Estimated Pooled Travel Cost Model

Ordinary least squares regression	Dep. Variable = Log VISITS
Observations = 305	Weights = ONE
Mean of LHS = 0.9244884E+01	Std. Dev of LHS = 0.1987143E+01
Std. Dev of residuals = 0.1438978E+01	Sum of squares = 0.6025611E+03
R-squared = 0.4980400E+00	Adj. R-squared = 0.4756157E+00
F [13, 291] = 0.2220981E+02	Prob value = 0.0000000E+00
Log-likelihood = 0.5366101E+03	Restr.(ß=0) Log-1 = -0.6417184E+03
Amemiya Pr. Criter. = 0.3610558E+01	Akaike Info. Crit. = 0.2165703E+01

ANOVA

Source	Variation	Degrees of Freedom	Mean Square
Regression	0.5978556E+03	13	0.4598889E+02
Residual	0.6025611E+03	291	0.2070657E+01
Total	0.1200417E+04	304	0.3948739E+01

Variable	Coefficient	Std.Error	t-ratio	Prob
Constant	8.3472	0.5370	15.545	0.00000
FARE	-0.56164E-02	0.1003E-02	-5.602	0.00000
GDP	0.73447E-04	0.1539E-04	4.772	0.00000
POP	0.13699E-08	0.5751E-09	2.382	0.01785
POPDEN	-0.20062E-03	0.6318E-04	-3.175	0.00166
BEACH	0.14725E-02	0.2569E-03	5.732	0.00000
PDAY	0.12633E-01	0.5436E-02	-2.324	0.02082
DIST	-0.76482E-04	0.5964E-04	-1.282	0.20069
TEMP	0.17252	0.4205E-01	4.103	0.00005
TEMPSQ	-0.29564E-02	0.1022E-02	-2.893	0.00410
PRECIP	-0.11087E-02	0.1342E-02	-0.826	0.40954
Q1	0.10804	0.2466	0.438	0.66162
Q2	-0.28269	0.2390	-1.183	0.23792
Q3	-0.13340	0.2448	-0.545	0.58617

Source: See text.

6. DISCUSSION

Using these results it is possible to examine the change in consumer surplus following a change in site attributes. The PTCM is also capable of answering the question "do British tourists have measurable use values for the low lying islands". The future facing these particular tourist destinations may not be a change in "site quality" but instead elimination through inundation under some climate change scenarios. These benefits are of course estimated only for British residents and many caveats apply, not least the assumptions made about the unimportance of the price and quality of substitute sites. The effect of this particular assumption is to make it appear that changes in the price and quality of alternative sites have no effect anywhere else whilst in reality of course, significant substitution between sites can be expected. Nevertheless, in the absence of anything better it is possible to use the model to predict the percentage change in the number of tourists visiting each country following a change in the attributes of the choice set as well as the ensuing change in consumer surplus and the overall worth of the site itself in terms of use values. These uses are illustrated for three different countries: Greece, Spain and the Seychelles. The first two are of particular interest since they are among the most popular tourist destinations for British people whilst the latter consists of a group of islands some of whose very existence is threatened by rising sea levels[6]. The impact of climate change on Greece and Spain is investigated inputting assumptions for the change in climate taken from the United Kingdom Meteorological Office's (UKMO) General Circulation Model as reported in the Houghton et al (1990). This model predicts a uniform increase of around 2°C for Southern Europe (30°N - 50°N) by the year 2030 complete with changes in seasonal precipitation patterns following "business as usual" emission assumptions.

Table 11.5 illustrating the situation for Greece indicates that there is a lengthening and a flattening of the tourist season with tourist numbers almost unchanged. The first, second and fourth quarters show an increase in consumer surplus whereas the third quarter marks a sharp decline as maximum daytime temperatures pass well beyond their optimum level of 29°C. Overall however there is a small increase in consumer surplus of just over £2.5 million. Table 11.6 illustrates the situation for Spain; a country whose attraction lies in its climate, low population density and many miles of beaches. The results of climate change for Spain are qualitatively similar to those for Greece, but given Spain's lower prices and slightly cooler climate the beneficial effects of climate change on tourism are more pronounced. There are large gains for both tourist numbers and consumer surplus in the first, second and fourth quarters. In the third quarter there is a small decline in tourist numbers but overall consumer surplus for trips to Spain increases by almost £55.5 million and the number of tourists visiting Spain goes up by more than 6%.

Finally, in Table 11.7 the total consumer surplus arising from trips to the Seychelles is estimated as being slightly more than £2 million pounds for quarters 1-3. This is the "use" value for this group of islands and the amount which would be lost if the whole group were inundated (as explained earlier this is a rather exaggerated proposition). In comparison with the gains from Spanish tourism this sum seems very small, as it would be for most of the island in the Indian and Pacific Oceans. The reason is that these islands are very small and quite distant (therefore expensive) and not visited much as a consequence. As a result they generate little consumer surplus. What this means is not that these low-lying islands are without value, but rather that their main value is likely to be in the form of existence rather than use values. The benefits from preserving these islands lie in the vicarious consumption of their services through films, literature and the appreciation of their cultural heritage. These services are not estimated through the travel cost technique.

Table 11.5. The Impact of Global Climate Change on British Tourism: the Effects of the UKMO's 2030 Scenario for Greece

Quarter	Temperature	Precipitation	Change in British tourists	Change in CS
1	+2°C	+5%	+16.0%	+£189,429
2	+2°C	-5%	+3.9%	+£3,992,970
3	+2°C	-15%	-2.8%	-£6,325,774
4	+2°C	-5%	+11.1%	+£4,691,700
Total			+0.7%	+£2,548,325

Table 11.6. The impact of Global Climate Change on British Tourism: the Effects of the UKMO's 2030 Scenario for Spain

Quarter	Temperature	Precipitation	Change in British tourists	Change in CS
1	+2°C	+5%	+19.4%	+£13,610,068
2	+2°C	-5%	+6.3%	+£16,541,888
3	+2°C	-15%	-0.5%	-£1,972,685
4	+2°C	-5%	+16.7%	+£27,258,392
Total			+6.3%	+£55,437,663

Source: See text.

Table 11.7. The Impact of Climate Change on British Tourism: the Effects of the
Inundation of the Seychelles

Quarter	Change in British tourists	Change in CS
1	-100 %	-£623,887
2	-100 %	-£1,204,152
3	100 %	-£191,938
4	100 %	n.a.
Total	-100 %	> -£2,019,977

Source: See text. Note that the change in consumer surplus is evaluated with unchanged temperatures.

7. CONCLUSIONS

It has been demonstrated that quarterly climate variables are able to explain differences in flows of tourists. In particular, it is shown that British tourists are attracted to climates which deviate little from an averaged daytime maximum of 29°C. Furthermore, as the attributes of low cost (i.e. nearby) destinations are likely to improve following climate change this is likely to result in a sizeable welfare gain to British tourists, even in the case of Southern European countries like Spain and Greece. Both these countries however, experience a lengthening and a flattening of the tourist season. In contrast, the losses experienced by the possible inundation of low-lying islands in the Indian and Pacific Oceans are likely to be small because these destinations are, at least to British residents, very expensive and consequently not much visited. But it is important to stress that the values which this paper seeks to estimate are use values and not total economic values which may be much greater.

At an empirical level there is also further work to be done in terms of specifying the demand equation: including alternative measures for precipitation (such as rain days), including variables representing hours of sunshine and variables representing the degree of personal security. It would be interesting to examine the role of socio-economic factors such as age and income in explaining travel patterns. At a theoretical level reallocation effects are not dealt with well in the PTCM in the sense that the effect of changes in the quality (and price) of substitute sites are set to zero. This is most unlikely to be a fair representation of what would in fact happen. In the RUM set-up by contrast, the number of holidays remains unchanged irrespective of the quality of the experience provided by different destinations so neither approach is entirely satisfactory.

NOTES

*   The Author would like to acknowledge helpful comments made on an earlier version of this paper by David Pearce, Kim Swales and Peter Pearson. The usual disclaimer applies.

1. In 1994 UK residents made almost 40 million trips abroad (CSO, 1995).

2. With the RUM an individual chooses from a set of alternatives according to the utility which they provide. The indirect utility function (V) associated with a particular site j and choice-occasion is:

$$V_j(z_j, y - p_j)$$

where zj is a vector of site attributes, y is income and pj is the price of visiting the site. The indirect utility function contains a random error term which means that the choices made cannot be predicted with certainty but only with a given probability. The random error term reflects the existence of unobserved site characteristics and/or variations in taste between individuals. An individual visits a particular site k provided that:

$$V_k(z_k, y - p_k) + e_k > V_j(z_j, y - p_j) + e_j$$

$$\forall \ sites \quad j \neq k$$

If the random error terms are distributed as type I extreme value variates then the probability of an individual i making choice j is given by:

$$Prob(Y_i = j) = \frac{e^{V_{ij}}}{\sum_{j=1}^{j=n} e^{V_{ij}}}$$

This is referred to as the Conditional Logit model. This model however has well known shortcomings of its own. One shortcoming the assumption of the Independence of Irrelevant Alternatives (IIA). A further defect of the RUM is that it is incapable of predicting any possible change in the total number of tourist trips made following a change in site attributes: all that is predicted is how an exogenously determined number of trips are allocated between different destinations (e.g. see Bockstael et al., 1986).

3. A computer program made available through the University of Michigan to measure the great circles distance between the world's capital cities can be found on the following internet site: http://www.indo.com/distance/.

4. The left-hand side variables are not transformed into logarithms since some of them take negative values.

5. The correlation coefficient is 0.81.

6. In fact the Seychelles consists of over 90 small islands situated in the Indian Ocean. They have a tropical climate and have recently become well known as a tourist resort. Most of the islands are low-lying but the largest island, Mahé, has hills rising to 3,000 ft. Hence the complete elimination of the Seychelles is unlikely.

REFERENCES

Bockstael N., McConnell K. and Strand I. (1991), "Recreation", in Braden, J. and C. Kolstadt (eds.), *Measuring the Demand for Environmental Quality*, Elsevier North Holland.

Caulkins P., Bishop R. and Bouwes N. (1986), "The Travel Cost Model for Lake Recreation: A Comparison of Two Methods for Incorporating Site Quality and Substitution Effects", *American Journal of Agricultural Economics*, 68: 291-297.

Central Statistical Office (1995), *Travel Trends*, HMSO, London.

Delft Hydraulics (1990), *Strategies for Adaptation to Sea Level Rise*, Report of the Coastal Management Zone Subgroup for WMO and UNEP, Geneva.

Freeman A. (1993), *The Measurement of Environmental and Resource Values*, Resources For the Future, Washington DC.

Houghton J., Jenkins G. and Ephraums J. (1990), *Climate Change: The IPCC Scientific Assessment*, Cambridge University Press, Cambridge.

Johansson P-O. (1987), *The Economic Theory and Measurement of Environmental Benefits*, Cambridge University Press, Cambridge.

Maddala G. (1977), *Econometrics*, McGraw-Hill, Singapore.

McConnell K. (1992), "On-Site Time in the Demand for Recreation", *American Journal of Agricultural Economics*, 74: 918-925.

Pearce E. and Smith C. (1994), *The World Weather Guide*, Helicon, Oxford.

Perry A. (1993), "Weather and Climate Information for the Package Holiday Maker", *Weather*, 48: 410-414.

Perry A. and Ashton S. (1994), "Recent Developments in the UK's Outbound Package Tourism Market", *Geography*, 79: 313-321.

Smith K. (1991), "Recreation and Tourism", in Parry M. (ed.), *The Potential Effects of Climate Change in the United Kingdom,* HMSO, London.

Smith V., Desvouges W. and Fisher A. (1986), "A Comparison of Direct and Indirect Methods for Estimating Environmental Benefits", *American Journal of Agricultural Economics*, 68: 280-290.

The Times World Atlas (1992) Times Books, London.

United Nations Development Programme (1995), *Human Development Report*, Oxford University Press, Oxford.

Wall G. (1992), "Tourism Alternatives in an Era of Global Climate Change", in Smith V. and Eadington W. (eds.) *Tourism Alternatives,* University of Pennsylvania Press, Philadelphia.

INDEX